Threatened animals
of Western Australia

by Andrew A Burbidge

GOVERNMENT OF WESTERN AUSTRALIA

DEPARTMENT OF
Conservation
AND LAND MANAGEMENT
Conserving the nature of WA

Above Noisy scrub-bird.

Photo – Babs and Bert Wells /CALM

Previous page Numbat.

Photo – Ken Stepnell/CALM

Publisher Department of Conservation and Land Management, 17 Dick Perry Ave, Kensington, Western Australia, 6151.

Managing editor Caris Bailey.

Editor Carolyn Thomson-Dans.

Design and production Maria Duthie, Alison Blackwell.

Cover photography (front) Masked boobies. Photo by Alex Steffe/Lochman Transparencies. **(back)** Western swamp tortoise. Photo by Babs and Bert Wells/CALM.

Author's acknowledgements This book could not have been written without help from many colleagues. I'd like to thank all those who contributed information (sometimes unpublished) and who reviewed draft chapters. My thanks go to (from the Department of Conservation and Land Management unless mentioned otherwise): Ian Abbott, John Blyth, Neil Burrows, Allan Burbidge, Val English, Bill Humphreys (Western Australian Museum), Ray Lawrie, Barbara York Main (Animal Sciences, The University of Western Australia), Nicky Marlow, Peter Mawson, Jalena May, Norm McKenzie, Brett Molony (Department of Fisheries), Keith Morris, David Pearson, Bob Prince, Dale Roberts (Animal Sciences, The University of Western Australia), Tony Start and Matt Williams.

Many people and organisations generously provided photographs and illustrations. My thanks go to AQWA, Kathie Atkinson, Australian Antarctic Division, C Baars, Stanley Breeden, Neil Burrows, ChevronTexaco Australia, Sarah Comer, Helen Crisp, Alan Danks, Paul Devine, Brad Durrant, Karen Edwards, Mick Eidam, Douglas Elford, Val English, O Ertok, Rosemary Gales, Stephanie Hill, Bill Humphreys, Darren Jew, Ken Johnson, Ron Johnstone, Greg Keighery, Peter Kendrick, Kevin Kenneally, Russell Lagdon, Don Langford (Bluesky Enterprises), Sheila Hamilton-Brown, Chris Hardwick, Mark Harvey, Terry Houston, Tricia Handasyde, Tony Howard, Edyta Jasinska, Darren Jew, G Johnstone, Jan van de Kam, Barbara York Main, Keith Morris, Peter Mawson, Justin McDonald, Jane McRae, David Morgan, Murdoch University, National Library of Australia, Shane O'Donoghue, David Pearson, Stuart Pearson, Adrian Priest, Bob Prince, Steve Pruett-Jones, Gordon Roberts, Shirley Slack-Smith, South Australian Museum, Steve Sturgeon, Martin Thompson, Ken Wallace, Rene Wanless, Grant Wardell-Johnson, Corey Whisson, Western Australian Museum, Buz Wilson, Eric Woehler (Polar-eyes.com) and Linda Worland. The inclusion of many photographs taken by Babs and Bert Wells, Michael James and Ken Stepnell (images all held in CALM's slide library) has made the book much more understandable and appealing.

The editing and design team from CALM's Strategic Affairs and Corporate Relations Division has done a great job in turning my words into a readable, attractive book. Special thanks go to Carolyn Thomson-Dans for her expert editing skills, as well as for her energy in driving the whole publication project, and to Maria Duthie for her superb design. Consultant Ali Blackwell assisted with the layout. Rhianna Mooney also provided valuable assistance. Ron Kawalilak encouraged me to prepare the book and Caris Bailey ensured resources were found to publish it. Neil Burrows and Keith Morris offered me the Post-retirement Research Fellowship that enabled me to use Departmental resources while writing this book. Merilyn Burbidge's support, constructive criticism and proofreading skills have been greatly appreciated.

Finally my thanks go to Keiran McNamara, Executive Director of the Department of Conservation and Land Management, for coauthoring the final chapter and working with me to develop a realistic vision for the conservation of Western Australia's amazing biodiversity, a vision that has the Department's backing.

ISBN 0 7307 5549 5

I dedicate this book to Emeritus Professor A R (Bert) Main, CBE, FAA.

It was Bert's inspirational lectures and supervision, and his dedication to science
and conservation, that led me into a career as a conservation biologist.
His guidance and example have been invaluable.

Contents

Introduction

Western Australia (WA) has many species of threatened animals—and, while efforts to bring back species from the brink of extinction are increasingly successful, the number is growing as our knowledge improves.

Biological diversity (biodiversity) is the variety of all life forms: the different plants, animals and microorganisms, their genes and the communities and ecosystems of which they are part. Biodiversity is usually recognised at three levels: genetic diversity, species diversity and ecosystem diversity.

'Genetic diversity' refers to the genetic differences within and between each species. Non-scientists best understand within-species diversity when recognising the differences between varieties of crop plants and breeds of livestock. Without genetic differences, breeders would not have been able to select for the characteristics they wanted. Chromosomes, genes and DNA—the building blocks of life—determine the uniqueness of each individual and each species. 'Species diversity' is measured most simply by the number of biologically defined species, and is easily understood by most people. So far, about 1.75 million species have been identified throughout the world, most of which are invertebrates such as insects. Most scientists think that there are probably around 13 million species, though estimates range from three million to 100 million. 'Ecosystem diversity' is the variety of ecosystems, such as those that occur in deserts, forests, wetlands, lakes, rivers and agricultural landscapes. It includes non-biological components such as soil, water and geological history, and other processes affecting them. In each ecosystem, living creatures—including people—form a community, interacting with one another and with the air, water and soil around them.

Australia is one of the most biologically-diverse countries in the world, with a large proportion of its species and most of its ecosystems found nowhere else. WA, comprising about a third of the Australian continent, accounts for a very high proportion of the nation's biodiversity. In particular, the south-west, from Shark Bay to the western edge of the Great Australian Bight, is one of only 25 megadiverse 'hotspots' in the world, only two of which are located in developed countries.

In recent years, the conservation of biodiversity has become a major objective of most countries in the world, and was adopted as an aim by the many nations (including Australia) that ratified the Convention on Biological Diversity (the Rio Convention) developed at the 1992 'Earth Summit' in Rio de Janeiro. The conservation of threatened species is the 'cutting edge' of biodiversity conservation. Threatened species can be likened to canaries in a coal mine—the loss of species from an ecosystem suggests that there is something seriously wrong with that system.

There are four main arguments for conserving living things and the ecosystems that they form. The first is that compassion demands their preservation. Compassion develops from the view that other species have a right to exist—the needs and desires of people should not be the only basis for ethical decisions.

The second argument is based on aesthetics. Plants and animals should be preserved because of their beauty, symbolic value or intrinsic interest. Most people would feel a loss if the world's beautiful and interesting plants and animals, and the wild places they inhabit, disappeared.

Compassion and a sense of wonder are two qualities that distinguish us from

Above Long-tailed dunnart. Although it was once considered to be rare and possibly threatened, research has shown this species to be relatively common and widespread, but restricted to a special habitat.

Opposite The orange migrant (*Catopsilia scylla*), which occurs in northern, eastern and central Australia, is just one of countless animal species that collectively contribute to Australia's biodiversity.

Photos – Babs and Bert Wells /CALM

Above Curlew sandpipers and red-necked stints at Roebuck Bay, Broome. This is one of the most biodiverse areas in Australia for birds, including these migratory shorebird species, as well as the invertebrates on which they feed, particularly those that inhabit the floor of the bay.

Photo – Jan van de Kam

other animals, but not all people share these attributes to the same degree. The third argument, however, is one that nearly everyone can understand—money. Our unique plants, animals and landscapes attract tourists to WA. Plants, animals and microorganisms provide all of our food and almost all of our medicines and drugs. They also provide renewable resources like paper, leather, fuel and building materials. So far, we have used only a minute proportion of nature's storehouse. Recent research is showing that many biological resources, including those from species many would regard as 'useless' today, will become valuable in the future. The new technology of genetic engineering depends on the variability of species and genes that occur in nature.

The fourth, and perhaps the most important, reason is that living things provide the indispensable life-support systems of our planet. They produce the oxygen we breathe, maintain the quality of the atmosphere, control climate, regulate freshwater supplies, generate and maintain the topsoil, dispose of wastes, generate and recycle nutrients, control pests and diseases, pollinate crops and provide a genetic store from which we can benefit in the future.

Increasing rates of extinction worldwide are part of a larger problem— we humans are getting out of balance with our environment. We have tended to regard the environment as limitless, and for 99 per cent of human history this view was justifiable. Now, our increasing population and improving technologies mean that we are able to assail the environment in ways it cannot sustain. Human attitudes to the conservation of biodiversity reflect our attitudes to the environment as a whole. If we continue to cause extinctions and the degradation of ecosystems then we will probably allow the biosphere to be damaged until it can no longer sustain us.

Once the challenge for humans was to conquer and subdue the environment. Now the challenge is to live sustainably in harmony with it. This will require continuing changes in attitudes and the development of special skills. It also requires that we understand the conservation biology of our State's species and ecological communities, and learn to manage our natural resources in an ecologically sustainable way.

Conserving WA's biodiversity is everyone's responsibility. However, the Western Australian Department of Conservation and Land Management has a special role to play, as the State Parliament has entrusted extremely valuable publicly-owned natural resources and natural areas to it. There are many techniques available to conserve biodiversity, but the identification and conservation of threatened species is a vital one.

What is a threatened species?

The Species Survival Commission of the World Conservation Union (also known as the IUCN—the acronym derived from its former and alternative name of the International Union for the Conservation of Nature and Natural Resources) has developed definitions and criteria to decide whether species should be included in its 'Red Lists'. This has not been an arbitrary process, but has involved much consultation with scientists and conservationists worldwide. The WA Threatened Species Scientific Committee and the Department of Conservation and Land Management have adopted the IUCN definitions and criteria as guides to placing species and subspecies on the State threatened species list. Thus, our lists are directly comparable to those of many other countries. IUCN Red List Category Definitions are provided below (note that in this context a taxon is a species, subspecies or variety).

The IUCN criteria used to decide whether a species is threatened, and if so to which category of threat it belongs, can be found at www.redlist.org/info /categories_criteria2001.html. The criteria are too detailed to publish in full here, but have been summarised on p. 4.

IUCN Red List Category Definitions

Extinct	A taxon is Extinct when there is no reasonable doubt that the last individual has died.
Extinct in the Wild	A taxon is Extinct in the Wild when it is known only to survive in cultivation, in captivity or as a naturalised population (or populations) well outside the past range.
Critically Endangered	A taxon is Critically Endangered when the best available evidence indicates that it meets any of the criteria A to E for Critically Endangered, and it is therefore considered to be facing an extremely high risk of extinction in the wild.
Endangered	A taxon is Endangered when the best available evidence indicates that it meets any of the criteria A to E for Endangered, and it is therefore considered to be facing a very high risk of extinction in the wild.
Vulnerable	A taxon is Vulnerable when the best available evidence indicates that it meets any of the criteria A to E for Vulnerable, and it is therefore considered to be facing a high risk of extinction in the wild.
Threatened	A collective term for taxa that are Critically Endangered, Endangered or Vulnerable.

IUCN Red List Categories and Criteria Version 3.1	Critically Endangered	Endangered	Vulnerable
A) Reduction in population size based on any of A1 – A4			
1) An observed, estimated, inferred or suspected population reduction of _____, over the last 10 years or 3 generations, whichever is the longer, where the causes are clearly reversible AND understood AND ceased, based on a, b, c, d or e.	≥90%	≥70%	≥50%
2) An observed, estimated, inferred or suspected population reduction of at least _____ over the last 10 years or 3 generations, whichever is the longer, where the reduction or its causes may not have ceased OR may not be understood OR may not be reversible based on a, b, c, d or e.	≥80%	≥50%	≥30%
3) A population size reduction of _____, projected or suspected to be met within the next 10 years or 3 generations, whichever is the longer (up to a maximum of 100 years) based on a, b, c, d or e.	≥80%	≥50%.	≥30%
4) An observed, estimated, inferred or suspected population reduction of _____ over any 10 year or 3 generation period, whichever is the longer (up to a maximum of 100 years in the future) where the time period must include both the past and the future, and where the reduction or its causes may not have ceased OR be understood OR may not be reversible, based on a, b, c, d or e.	≥80%	≥50%	≥30%
a) Direct observation. b) An index of abundance appropriate for the taxon. c) A decline in area of occupancy, extent of occurrence and/or quality of habitat. d) Actual or potential levels of exploitation. e) The effects of introduced taxa, hybridisation, pathogens, pollutants, competitors or parasites.			
B) Geographic range in the form of either B1 OR B2			
1) Extent of occurrence _____ and estimates indicating at least 2 of a-c	<100 km²	<5000 km²	<20 000 km²

2) Area of occupancy _____ and estimates indicating at least 2 of a-c
a) Severely fragmented or known to exist at no more than _____ locations.
b) Continuing decline, observed, inferred or projected, in ANY of the following:
  (i) extent of occurrence, (ii) area of occupancy, (iii) area, extent and/or quality of habitat, (iv) number of locations or subpopulations, (v) number of mature individuals.
c) Extreme fluctuations in any of the following:
  (i) extent of occurrence, (ii) area of occupancy, (iii) area, extent and/or quality of habitat, (iii) number of locations or subpopulations, (iv) number of mature individuals.
	<10 km²
	
	one
		500 km²
		
		five
			<2000 km²
			
			ten

C) Population estimated to number _____ mature individuals and either:	<250	<2500	<10 000
1) an estimated continuing decline of at least _____ within 3 years or one generation whichever is the longer (up to a maximum of 100 years in the future) OR	25%	20%	10%

2) a continuing decline, observed, projected, or inferred, in numbers of mature individuals AND at least one of a-b
a) population structure in the form of one of:
  (i) no subpopulation estimated to contain more than _____ mature individuals; OR
  (ii) at least 90% of mature individuals in one subpopulation.
b) extreme fluctuations in the number of mature individuals.
	50	250	1000

D) (CR and EN) Population size estimated to be less than _____ mature individuals	50	250	not applicable
D) (VU ONLY) Population very small or restricted in the form of either:			
1) population estimated to number less than _____ mature individuals; OR	not applicable	not applicable	1000
2) population with a very restricted area of occupancy (typically less than 20 km²) OR number of locations (typically five or fewer) such that it is prone to the effects of human activities or conjectural events within a very short period of time in an uncertain future, and is thus capable of becoming Critically Endangered or even Extinct in a very short time period.	not applicable	not applicable	applies
E) Quantitative analysis showing probability of extinction in the wild is at least _____	50% within 10 years or 3 generations, whichever is the longer (up to a maximum of 100 years)	20% within 20 years or 5 generations, whichever is the longer (up to a maximum of 100 years)	10% within 100 years

How many threatened animals are there in WA?

In WA, listing of threatened species is guided by the Department of Conservation and Land Management's Policy Statement 'Conserving threatened species and ecological communities'.

At April 2003, 185 animals were listed as threatened in WA (see Table 1). Most listed species are vertebrates, of which the largest group is birds. This reflects our knowledge of the different groups. The small number of invertebrates—which comprise at least 95 per cent of all living organisms—is due to a lack of knowledge, and many more species may in fact be extinct or threatened than the official list indicates.

Making comparisons between WA and other Australian States is not straightforward. Most other States list species according to their status within that State, whereas WA lists species according to their national status. This means that some States, such as New South Wales, have listed species as threatened that, while rare within that State, are not threatened nationally. However, comparisons can be made using a list agreed in 2002 between threatened fauna experts in the nature conservation agency in each State, the Northern Territory and the Australian Capital Territory (at the time of writing, the list developed under the *Commonwealth Environment Protection and Biodiversity Conservation Act 1999* had not been updated to reflect many recent changes agreed to by the States).

Table 1   Number of listed Extinct, Extinct in the Wild and Threatened animals (species and subspecies) in WA in April 2003.

	EX	EW	Sub-total	CR	EN	VU	Threatened
mammals (Mammalia)	10	1	11	2	8	31	41
birds (Aves)	2		2	2	8	32	42
reptiles (Reptilia)				1	2	14	17
frogs (Amphibia)				1		2	3
sharks (Chondrichthyes)						2	2
bony fish (Actinopterygii)						2	2
insects (Insecta)	1		1		3		3
millipedes (Diplopoda)				1	1	3	5
spiders, etc (Arachnida)				3	7	8	18
crustaceans (Crustacea)				6	1	11	18
bristleworms (Polychaeta)				1			1
snails (Gastropoda)	4		4	21	5	7	33
TOTAL	17	1	18	38	35	112	185

CR Critically Endangered; EN Endangered; VU Vulnerable; EX Extinct; EW Extinct in the Wild

Table 2 Number of species and subspecies of threatened fauna in Australia and each State*, the Northern Territory, External Territories and Australia's territorial seas, as at June 2002.

	Red List Category	Australia	WA	SA	VIC	NSW	TAS	QLD	NT	External Territories#
MAMMALS	EX	27	14	17	5	10	1	6	9	2
	EW	2	1	2					1	
	CR	7	1		1			4	1	1
	EN	24	9	4	5	4	1	11	5	1
	VU	51	30	16	9	11	6	14	13	4
Threatened mammals		82	40	20	15	15	7	29	19	6
BIRDS	EX	27	2	1		10	5	1	1	9
	CR	33	4	6	5	5	10	8	1	18
	EN	40	10	14	14	11	12	8	4	11
	VU	84	34	26	31	36	34	9	7	36
Threatened birds		157	48	46	50	52	56	25	12	65
REPTILES	EX	1						1		
	CR	2	1		1					
	EN	8	1	1	1	2	1	4		
	VU	37	13	3	6	10		16	9	7
Threatened reptiles		47	15	4	8	12	1	20	9	7
FROGS	EX	4						4		
	CR	8	1		2	4		3		
	EN	8		1	1	2		7		
	VU	10	2		4	7		2		
Threatened frogs		26	3	1	7	13	0	12	0	0
SHARKS, SKATES, RAYS ETC	EX									
	CR									
	EN									
	VU	2	2	2	2	2	2			1
Threatened sharks		2	2	2	2	2	2	0	0	1
BONY FISH	EX									
	CR	4			2	1	2			
	EN	11		1	2	3	3	4		
	VU	2	2	2	2	2	2			1
Threatened bony fish		17	2	3	6	6	7	4	0	1
EARTHWORMS	EX									
	CR									
	EN	1					1			
	VU	1			1					
Threatened earthworms		2	0	0	1	0	1	0	0	0
SEA STARS	EX									
	CR									
	EN	2					2			
	VU									
Threatened sea stars		2	0	0	0	0	2	0	0	0
INSECTS	EX	5	1				3			1
	CR	3			1	1	1			
	EN	12	3		3	3	4			
	VU	16			5		6		5	
Threatened insects		31	3	0	9	4	11	0	5	0
MILLIPEDES	EX									
	CR	1	1							
	EN									
	VU	3	3							
Threatened millipedes		4	4	0	0	0	0	0	0	0

	Red List Category	Australia	WA	SA	VIC	NSW	TAS	QLD	NT	External Territories#
SPIDERS AND SCORPIONS	EX	1					1			
	CR	4	3				1			
	EN	7	6						1	
	VU	9	9							
Threatened arachnids		20	18	0	0	0	1	0	1	0
CRUSTACEANS	EX									
	CR	7	6		1					
	EN	4	1		2		1			
	VU	17	10		5	1	2			
Threatened crustaceans		28	17	0	8	1	3	0	0	0
VELVET WORMS	EX									
	CR									
	EN	1					1			
	VU									
Threatened velvet worms		1	0	0	0	0	1	0	0	0
SNAILS AND OTHER GASTROPODS	EX									
	CR									
	EN	5	1			3	3			
	VU	5	1		2		2			
Threatened snails		10	2	0	2	3	5	0	0	0
TOTAL EXTINCT + EW		67	18	20	5	20	10	12	11	12
TOTAL THREATENED		429	154	76	108	108	97	90	46	80

Table 2 shows the number of extinct and threatened animal species in each State, as well as the Northern Territory and Australia's external territories and territorial seas. Differences in numbers of WA-listed animals between Tables 1 and 2 are due to changes in listings that occurred in 2003, when a significant number of invertebrates (mainly terrestrial molluscs) were listed, as well as some other changes.

In making comparisons, it needs to be noted that there is much more information on some animal groups than others. Birds are the best known of all animal groups, and *The action plan for Australian birds 2000* allocated conservation status to many subspecies. At the other end of the spectrum, most invertebrate groups are very poorly known and, while some States have made a greater effort to list invertebrates than others, listing is very incomplete. Additional research into the conservation status of invertebrates is urgently needed.

Some differences between WA and other States and the Northern Territory are evident. WA, the largest Australian State, covers three major biogeographic regions: the south-west; arid and desert areas; and the Kimberley. Many animal species are endemic to the south-west (they occur nowhere else) and some of these, such as the ngilkait (Gilbert's potoroo) and the western swamp tortoise, are threatened. Other centres of endemism occur, for example, in the north Kimberley, Pilbara and central ranges. Some species, especially some mammals, can be termed 'new endemics'. These species once had more widespread distributions, but are now restricted to WA. They include the woylie (*Bettongia penicillata*), boodie (*B. lesueur*) and numbat (*Myrmecobius fasciatus*), although all three of these species are now being reintroduced to other States.

Because of its large size relative to other States, and because of its special features, WA has a high proportion of extinct and threatened animals. Table 2 shows that 27% of Australian Extinct and Extinct in the Wild fauna occurred in WA and 36% of Australia's threatened fauna occur here, a higher proportion than any other State. The number of known extinctions is also relatively high, but not as high as in South Australia and New South Wales.

* Includes oceanic birds occurring in State and Territory waters.

\# External territories (islands) and territorial seas. ACT data incorporated into NSW.

International comparisons

The 2002 Red List of Threatened Animals allows us to make comparisons between Australia and other countries. The information in the list is derived from a wide variety of sources, and may in some cases be inaccurate or incomplete, but is the best available. Table 3, derived from the 2002 IUCN Red List, shows the number of animals listed as Threatened, Extinct, and Extinct in the Wild throughout the world. Birds have the highest number, followed closely by mammals and then by gastropod molluscs. These data reflect the state of knowledge of the different taxonomic groups, rather than actual numbers. Birds, mammals and gastropods are the best-known groups, the last of these because of the large number of shell collectors.

Table 3 Number of species of threatened animals in the world in 2002 IUCN Red List (see www.redlist.org).

Class *	EX	EW	Sub-total	CR	EN	VU	Sub-total
mammals	74	3	77	181	339	617	1137
birds	129	3	132	182	326	684	1192
reptiles	21	2	23	55	79	159	293
amphibians	7	0	7	30	37	90	157
lampreys	1	0	1	0	1	2	3
sharks, skates, rays etc	0	0	0	4	16	19	39
bony fish	80	11	91	152	126	421	699
coelacanth	0	0	0	1	0	0	1
sea urchins, starfish etc	0	0	0	0	0	0	0
spiders, scorpions etc	0	0	0	0	1	9	10
centipedes	0	0	0	0	0	1	1
crustaceans	7	1	8	56	73	280	409
insects	71	1	72	46	118	393	557
horseshoe crabs	0	0	0	0	0	0	0
velvet worms	3	0	3	1	3	2	6
leeches	0	0	0	0	0	0	0
earthworms	0	0	0	1	0	4	5
marine bristle worms	0	0	0	1	0	0	1
bivalves	31	0	31	52	28	12	92
snails etc	261	12	273	170	208	469	847
nemertine worms	0	0	0	0	0	2	2
flatworms	1	0	1	0	0	0	0
corals and their relatives	0	0	0	0	0	2	2
TOTAL	686	33	719	932	1355	3166	5453

*Mammals (Mammalia), birds (Aves), reptiles (Reptilia), amphibians (Amphibia), lampreys (Cephalaspidomorphi), sharks, skates, rays and chimaeras (Elasmobranchii), bony fish (Actinopterygii), coelacanth (Sarcopterygii), sea stars, urchins and sea cucumbers (Echinoidea), spiders, scorpions and other arachnids (Arachnida), centipedes (Chilopoda), crustaceans (Crustacea), insects (Insecta), horseshoe crabs (Merostomata), velvet worms (Onychopora), leeches (Hirudinoidea), earthworms (Oligochaeta), marine bristle worms (Polychaeta), bivalve molluscs (Bivalvia), snails etc (Gastropoda), nemertine worms (Enopla), flatworms (Turbellaria), corals, sea anemones and other anthozoans (Anthozoa).

CR Critically Endangered; EN Endangered; VU Vulnerable; EX Extinct; EW Extinct in the Wild

Table 4 2002 IUCN Red List animal data from selected countries (for the full list see www.redlist.org)

Country	EX	EW	Sub-total	CR	EN	VU	Sub-total
United States of America	236	3	239	157	164	510	831
Australia	35	0	35	49	119	331	499
Indonesia	3	0	3	44	102	242	388
Brazil	5	1	6	49	70	155	274
Mexico	21	6	27	56	89	114	259
South Africa	9	0	9	27	48	164	239
China	3	0	3	26	67	128	221
India	0	0	0	18	52	150	220
Japan	14	0	14	19	64	67	150
Tanzania	0	1	1	12	68	64	144
Thailand	1	1	2	10	34	71	115
New Zealand	21	0	21	9	20	75	104
France	2	0	2	8	7	83	98
French Polynesia	67	9	76	25	16	20	61
Austria	2	0	2	9	13	39	61
Germany	1	0	1	1	5	47	53
Mauritius	41	1	42	8	15	26	49
United Kingdom	1	0	1	2	5	21	28
Netherlands	0	0	0	1	1	21	23
Ireland	1	0	1	0	2	6	8
Iceland	1	0	1	0	4	2	6

According to the IUCN 2002 Red List, Australia has the second largest number of threatened species, with the United States having the highest number of both extinct species and threatened species. Once again, however, these figures reflect knowledge in these countries, rather than absolute totals: compared to Australia, the United States has better quality data on the conservation status of many of its invertebrate animal groups, and shows correspondingly high totals of threatened species in these groups. Many countries have very limited information on the conservation status of their threatened animals. However, when comparisons are made between Australia and some European countries with well-known faunas, and restricted to well-known groups such as mammals and birds, it is evident that many more Australian animals have become extinct and are threatened with extinction than in most other developed countries.

Table 5 Numbers of threatened animal species in selected classes in IUCN 2002 Red List, Australia and WA.

Group	Red List	Australia	WA
mammals	1137	82 (7.2%)	40 (3.5%)
birds	1192	157 (13.2%)	48 (4.0%)
reptiles	293	47 (16.0%)	15 (5.1%)
amphibians	157	26 (16.6%)	3 (1.9%)
sharks, rays etc	39	2 (5.1%)	2 (5.1%)
bony fish	699	17 (2.4%)	2 (0.3%)
arachnids	10	20*	18*
crustaceans	409	28 (6.8%)	20 (4.9%)
insects	557	31 (5.6)	20 (3.6)
gastropods	847	10 (1.2%)	2 (0.2%)

Note: 31 more gastropods were listed as threatened in WA in 2003
* Data for arachnids in IUCN Red List are out of date

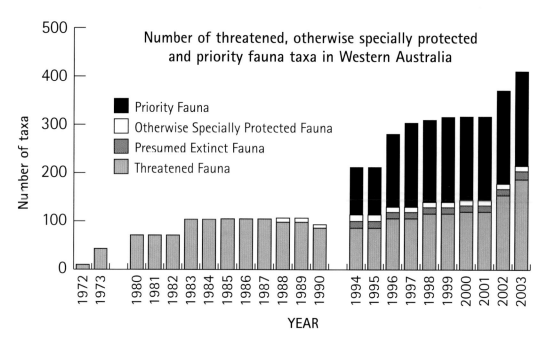

Number of threatened, otherwise specially protected and priority fauna taxa in Western Australia

Priority Fauna
Otherwise Specially Protected Fauna
Presumed Extinct Fauna
Threatened Fauna

Comparing the number of threatened animals in Australia with the world totals in the Red List (Table 5) indicates that WA has a relatively high proportion of the world's threatened mammals (3.5%), birds (4.0%), reptiles (5.1%), sharks (5.1%) and crustaceans (4.9%). However, as data in the Red List are derived from a variety of sources and are incomplete, the figures should be treated with caution.

While there are relatively high quality data for WA on birds and mammals, data on many other groups, especially invertebrates, are often insufficient to be reasonably sure that a species meets criteria for threatened status. For many groups of animals, data are virtually non-existent.

To help overcome this problem, the Department of Conservation and Land Management maintains a 'Priority Fauna List' as well as the legal list of threatened animals. The Priority list has no legal status, but helps the Department target surveys towards species most likely to be at risk, and assists it and other government agencies responsible for planning and environmental impact assessment. The Priority list includes—but is not limited to—species that meet the IUCN Red List category 'near threatened'. Priority code definitions are on p. 11.

At April 2003, 196 species or subspecies were on the WA Priority Fauna List (Table 6). Reptiles accounted for 54 of these, as many reptile species are known from very few specimens from highly restricted geographic ranges. As knowledge improves, some of these will prove not to be threatened, but others may be added to the list of threatened fauna.

The number of listed threatened animals and Priority Fauna has increased over the years (see above graph)—mainly due to improved knowledge, although some species have become threatened due to habitat loss and other factors. Chapter 2 discusses the reasons for species becoming extinct or threatened.

Table 6 Number of species and subspecies in the WA 'Priority Fauna List', April 2003

Class	P1	P2	P3	P4	P5	Total
mammals	1	2	4	16	3	26
birds		2	4	23		29
reptiles	20	25	4	5		54
amphibians	2			1		3
lampreys and hagfish	1					1
bony fish	2	10	1	3		16
insects	7	7	6	1		21
arachnids	1	3		1		5
crustaceans	9	3		2		14
gastropod molluscs	14	8	4			26
bivalve molluscs				1		1
TOTAL	57	60	23	53	3	196

See p. 11 for explanation of Priority codes. See Table 3 on p. 8 for description of Classes.

Possibly threatened species that do not meet survey criteria are added to CALM's Priority Fauna Lists under Priorities One, Two or Three. These three categories are ranked in order of priority for survey and evaluation of conservation status, so that consideration can be given to their declaration as threatened fauna. Species that are adequately known, or rare but not threatened, or meet criteria for Near Threatened, or that have been recently removed from the threatened list for other than taxonomic reasons, are placed in Priority Four. These species require regular monitoring. Conservation Dependent species are placed in Priority Five.

Priority One: Poorly-known species

Species that are known from one or a few collections or sight records (generally less than five), all on lands not managed for conservation (such as agricultural or pastoral lands, urban areas, Shire, Westrail and Main Roads road, gravel and soil reserves, and active mineral leases) and under threat of habitat destruction or degradation. Species may be included if they are comparatively well-known from one or more localities, but do not meet adequacy of survey requirements and appear to be under immediate threat from known threatening processes.

Priority Two: Poorly-known species

Species that are known from one or a few collections or sight records (generally less than five), some of which are on lands not under imminent threat of habitat destruction or degradation (such as national parks, conservation parks, nature reserves, State forest, vacant Crown land and water reserves). Species may be included if they are comparatively well-known from one or more localities, but do not meet adequacy of survey requirements and appear to be under threat from known threatening processes.

Priority Three: Poorly-known species

Species that are known from collections or sight records from several localities not under imminent threat, or from few but widespread localities with either large population sizes or significant remaining areas of apparently suitable habitat, much of it not under imminent threat. Species may be included if they are comparatively well-known from several localities but do not meet adequacy of survey requirements and known threatening processes exist that could affect them.

Priority Four: Rare, Near Threatened and other species in need of monitoring

(a) Rare. Species that are considered to have been adequately surveyed, or for which sufficient knowledge is available, and that are considered not currently threatened or in need of special protection, but could be if present circumstances change. These species are usually represented on conservation lands.

(b) Near Threatened. Species that are considered to have been adequately surveyed and that do not qualify for Conservation Dependent, but that are close to qualifying for Vulnerable.

(c) Species that have been removed from the list of threatened species during the past five years for reasons other than taxonomy.

Priority Five: Conservation Dependent species

Species that are not threatened but are subject to a specific conservation program, the cessation of which would result in the species becoming threatened within five years.

References

IUCN (1994). *Red List Categories and Criteria.* Prepared by the Species Survival Commission. IUCN: Gland.

IUCN (2000). *IUCN Red List Categories version 3.1.* Prepared by the Species Survival Commission. IUCN: Gland, Switzerland and Cambridge, UK. [www.iucn.org]

IUCN (2002). *The 2002 Red List of threatened species.* IUCN, Gland. [www.redlist.org]

Mittermeier, R.A., Myers, N. and Mittermeier, C.G. (1999). *Hotspots. Earth's biologically richest and most endangered terrestrial ecoregions.* Cemex and Conservation International, Mexico City.

Threatening processes

Western Australia has many threatened animals—more than most countries in Europe—a number that is increasing. What has led to this situation?

Activities and agents that cause species to become threatened or extinct are called 'threatening processes'. In WA, three general threatening processes—habitat destruction, habitat degradation and invasive (introduced) species—have been major contributors to the decline and extinction of species. To these three horsemen of the apocalypse must now be added a fourth—climate change. Other processes, such as over-exploitation, will also be discussed.

Habitat destruction

Land clearing, mainly for agriculture but also for urban and other developments, has been a major cause of species becoming extinct or threatened with extinction in Australia. If an animal's habitat is destroyed, the animal cannot survive. While a few animals have benefited from clearing, and the resulting expansion of habitats to which they are adapted (crested pigeons, galahs and ravens are good examples), most species simply disappear from cleared areas. In recent years, our society has tried to balance habitat destruction with the creation of conservation reserves (protected areas)—national parks, nature reserves and other reserves—and by promoting multiple use, for example, in State forest. Unfortunately, however, the acceleration of agricultural land clearing after the First World War, and particularly after the Second World War, was not accompanied by significant habitat reservation. Although some important parks and reserves were created before the 1960s, moves to develop a 'comprehensive, adequate and representative' (CAR) reserves system did not really get underway until after most land clearing had taken place. In some parts of the State, notably the agricultural areas of the south-west, it is now impossible to create a CAR reserves system.

Some land has been left uncleared in agricultural regions, but most bush remnants (including most conservation reserves) are very small and isolated. Habitat fragmentation, as this process is called, leads to the gradual loss of species. This is partly because remnants are too small for sufficient numbers of an animal to survive, partly because the intervening gaps are either too large or too inhospitable for animals to move across them between remnants, and partly because remnants, particularly the smaller ones, degrade. Recent overseas research suggests that disease is a major factor that leads to the loss of species from remnants. In agricultural areas, the Department of Conservation and Land Management has the difficult task of managing many small nature reserves—the median size of the 640 reserves in the Wheatbelt region is only 112.5 ha. Much larger reserves would be desirable, but large areas of uncleared and unreserved bushland no longer exist.

Habitat degradation

Even where habitat is uncleared, it can be made unsuitable for wildlife in a number of ways. Apart from damage caused by introduced animals and diseases (discussed below), changes may be due to altered hydrology, unsuitable fire regimes, grazing, chemical drift, weed invasion and increased incidence of naturally-occurring disease.

Salinity and waterlogging

Removal of native vegetation for farming or other land uses leads to dramatic changes in an area's hydrology. In much of the south-west of WA, the most obvious effect of changed hydrology is salinisation (increased surface salinity). In most agricultural areas, the natural water table consists of a narrow band of

Above The dibbler is threatened by feral cats and inappropriate fire regimes. Island populations are threatened by the potential introduction of predators, including rats.

Opposite The introduction of the European red fox to Australia led to a conservation tragedy of enormous dimensions.

Photos – Babs and Bert Wells /CALM

fresh water above a broader layer of salt water, the salt having accumulated from rainfall over many thousands of years. Removal of deep-rooted perennial vegetation such as trees and shrubs, and replacement with shallow-rooted annuals such as cereal crops and pasture, causes the water table to rise. When the salt water reaches the surface, it kills the vegetation. This is not just a local problem: salt water tables rise throughout whole catchments and may result in the demise of native plants in vegetation remnants, as well as agricultural crops.

Salinity is now one of the most significant threats to biodiversity in the south-west of WA. Recent detailed biogeographic research by Department of Conservation and Land Management scientists suggests that hundreds of species of native plants and animals will become extinct, unless salinity can be effectively and quickly controlled.

Changed fire regimes

'Fire regime' is a collective term embracing fire intensity, size, season and frequency. Fire ignited by lightning has occurred naturally for millions of years, and people have used fire in Australia for tens of thousands of years. Since European settlement less than 200 years ago, the fire regime in many parts of WA has changed dramatically. Aboriginal people used fire extensively to manage the landscape and their food resources, and for hunting, signalling, warmth and to open up the country to make travel easier. European settlers had very different attitudes to fire, as they had fixed assets such as buildings, fences and developed pastures to protect. They also had a poor understanding of fire as a natural part of land management in Australia, because fire in most parts of Europe is not a significant part of the natural cycle of the landscape as it is

Below The now-saline Lake Dumbleyung in Western Australia's Wheatbelt region. Salinity is one of the most significant threats to biodiversity in the south-west of Western Australia.

Photo – Michael James/CALM

here. In many parts of WA, the fire regime has changed from one of fairly frequent, mostly small, fires during much of the year to occasional large summer fires. This is particularly evident in the spinifex (*Triodia*) dominated hummock grasslands of the deserts and the Pilbara, and in the tropical savannahs of the Kimberley. On the other hand, fire frequency in most rangelands has reduced as a result of reduction in fuel due to stock grazing. In the forests of the south-west, fire is still a controversial issue, with opinions on the best regime varying from no planned fires to frequent fuel-reduction prescribed burning to prevent large wildfires from developing. Away from the urban and closely-settled south-west, fires are less managed and are probably doing far more environmental harm than in the forests. However, in these remote areas, public interest is much lower, little research is undertaken and resources to manage biodiversity and fire are scarce.

Studying the effects of different fire regimes on wildlife is difficult, as the effects of a fire vary according to the time that has elapsed since the last fire or fires, the fire intensity and season, the extent of the fire, the climate, and rainfall before and after the fire. Studying the effects of a single fire, especially when there is little information from prior years, is of limited value. Thus, there is an urgent need for long-term research into fire regimes in all major ecological communities in the State. Managing fire regimes on conservation lands is very complex and requires much more scientific research and management resources than are currently provided.

Different species of animals may be adapted to different fire regimes. Some research suggests that no single fire regime is optimal for all fauna of an area. It is possible, therefore, for animals to become rarer because of either too many or too few fires, or because fires are too large.

Above An Aboriginal hunting fire in the western desert. In the Western Australian deserts, the fire regime has changed from one of frequent small fires to large infrequent summer wildfires. These changes may have contributed to the disappearance of many animal species.

Photo – Andrew Burbidge /CALM

Some animals, such as some native rodents and large kangaroos, tend to be most abundant a few years after a fire has burnt their habitat, then become less common as the regenerating vegetation thickens. Other animals do best in long-unburnt places, as they are unable to recolonise a burnt area until it has developed dense vegetation. Still others require climax vegetation, and will decline as plants become over-mature. While much additional research is needed to clarify the best fire regimes for many of WA's threatened animals, current information suggests that some are adapted to relatively long-unburnt or patchily-burnt areas and that extensive, hot fires are detrimental to them, particularly if all or most of a vegetation remnant is burnt out. Examples include the malleefowl, noisy scrub-bird, dibbler and ngilkait (Gilbert's potoroo). On the other hand, the sandhill dunnart in the Great Victoria Desert lives in climax spinifex (*Triodia*) hummock grassland, but disappears once the spinifex starts to get old and hummocks start to die in the middle.

Many of the locally or totally extinct medium-sized mammals of the arid interior disappeared as the fire regime changed when Aboriginal people abandoned their traditional nomadic lifestyle for settlements. However, the introduction of foxes and cats also contributed to this massive extinction, as foxes became common in the deserts at about the same time that burning practices changed (see below and Chapter 5). However, it is interesting that many extinct desert mammals remained longest in the northern Gibson Desert and southern Great Sandy Desert, country occupied by Pintupi people until the 1950s and 1960s. There, mammals probably survived because Aboriginal people continued to manage the land traditionally (lighting many, mostly small fires to maintain a fine-grained mosaic of vegetation of different ages since the last fire) and hunted feral cats for food. Dingoes remained common in such areas, probably suppressing fox numbers.

Species that are restricted to a particular fire-sensitive habitat may be very susceptible to changed fire regimes.

Below A black rat (*Rattus rattus*).

Photo – Babs and Bert Wells /CALM

Many species of land snails in the family Camaenidae in the Kimberley are restricted to very few 'rainforest' patches (usually small areas of monsoonal vine thicket) and today's frequent, hotter, late-dry-season fires are rapidly destroying these areas. The changes due to fire are exacerbated by the presence of cattle (stock or feral) and feral donkeys, which move into recently-burnt vine thickets to use them for shelter and food, accelerating the rate of degradation. There is increasing evidence that the mammal fauna of the north-west Kimberley, thought to be intact, is declining because of frequent large fires.

Other processes causing habitat degradation

Other major processes leading to habitat degradation include weed invasion and grazing by introduced animals. There are many other processes that may be significant in some places. For example, the edges of remnants may be affected by chemical (such as fertiliser or herbicide) drift from adjacent lands, reserves are affected by rubbish dumping and off-road vehicle use, and wetlands (including estuaries) are suffering eutrophication from fertiliser run off.

Invasive species

As we saw in Chapter 1, statistics showing the number of animal extinctions in different countries demonstrate that species restricted to islands are particularly susceptible to extinction due to invasive species. Australia, the world's largest island, is not immune, and has been greatly affected by invasive species: predators, herbivores, disease organisms and weeds.

Invasive species are not limited to terrestrial environments. Several species of marine organisms have established in Australian waters, having arrived mainly in the ballast water of ships. Some, such as the northern Pacific sea star (*Asterias*

amurensis), which is a voracious predator of a wide variety of native animals, have the potential to eliminate native species. This sea star has already almost exterminated the Tasmanian spotted handfish (*Brachionichthys hirsutus*) by eating its egg masses.

Predators

Three recently-introduced predators—the European red fox (*Vulpes vulpes*), the feral cat (*Felis sylvestris catus*) and the black rat (*Rattus rattus*)—have had a major impact on terrestrial native species. *Western Shield*, a major long-term project being carried out by the Department of Conservation and Land Management, aims to reduce the impact of these predators and, where possible, reintroduce and introduce threatened species that they have eliminated from conservation lands (see Chapter 3).

Foxes were released in Victoria during the 1860s and 1870s for 'sporting' purposes. Once established, they spread very rapidly both north and west. They were first reported in the south-west of WA in the 1920s, and were common by the 1930s. They are best adapted to temperate and subtropical areas and do not fare well in the hot, wet tropics, nor in very rocky areas. Thus they are either absent or present only in low numbers in far northern deserts, and are not established in the Pilbara uplands or in the Kimberley—although individual animals are found in some years. Research by Department of Conservation and Land Management scientists has demonstrated that foxes are major predators of many native animals and can cause species to become extinct (see Chapter 3).

Feral cats became established in WA well before foxes. It has been suggested that they established before European settlement from early shipwrecks, but available evidence does not support this. They certainly became wild soon after 1788 in eastern Australia,

and then spread rapidly throughout the continent. Because of the lack of an effective means of broadscale feral cat control, it has not been possible to conduct scientific experiments to document the effect of cats on native mammals. However, evidence from islands indicates that they have been responsible for eliminating native mammals from arid areas, and historical evidence suggests that, in the south-west of WA, they caused the extinction of several species that disappeared before the arrival of the fox. They have also been responsible for the failure of a number of attempted mammal reintroductions in the arid zone.

Black rats and brown rats, or Norway rats (*Rattus norvegicus*), arrived in Australia with early sailing boats. In WA, brown rats have been reported only near Perth. Black rats, on the other hand, have adapted very well to our conditions. They are common in bushland areas in the south-west (as far inland as Frank Hann National Park) and along the west and north-west coasts, including the North West Cape peninsula, as well as near Broome, Fitzroy Crossing and Kununurra. Black rats were introduced to many northern WA islands by pearling boats in the second half of the nineteenth century. Most of these populations have now been eradicated. Black rat control on the mainland has not been widespread, but has been necessary in one of the nature reserves set aside to conserve the Critically Endangered western swamp tortoise.

Above The western swamp tortoise, threatened by land clearing, foxes, feral cats, black rats and changing hydrology.

Photo – Babs and Bert Wells /CALM

A third species of invasive rat, the Polynesian rat or Pacific rat (*Rattus exulans*), has been found on Adele and Sunday islands, off the Kimberley coast, and may be more widespread than we think. The house mouse (*Mus domesticus*) now occurs in most of Australia. Its effects on native wildlife are poorly understood, but its ability to outbreed small native mammals—and its exceptionally varied and adaptable diet—suggests that it may have a detrimental effect through competition for food.

Above The introduction of the dingo to Australia less than 4000 years ago led to the disappearance of native animal species, such as the thylacine, from the mainland.

Photo – Babs and Bert Wells /CALM

Another mammal predator that was introduced to Australia some time ago has had, and may still be having, a major effect on native wildlife. The dingo (*Canis lupus dingo*), a subspecies of wolf, was introduced to Australia from Asia about 3500 to 4000 years ago. It is generally accepted that dingoes exterminated the thylacine (*Thylacinus cynocephalus*) and Tasmanian devil (*Sarcophilus harrisii*) from mainland Australia—these species survived in Tasmania, where dingoes did not occur. Aboriginal people introduced dingoes to some northern Australian islands (including large Kimberley islands such as Augustus, Bigge, Uwins and Middle Osborne), but were not able to take them to most of WA's very valuable (from a biodiversity conservation viewpoint) islands, such as Barrow, Bernier and

Dorre. Today, many mainland populations of dingoes include hybrids with domestic dogs, which were bred by people from wolves.

Yet another mammal predator of concern is the ferret, the domesticated form of the polecat (*Mustela putorius*). It has become feral in Tasmania and New Zealand, and occasionally turns up in the wild in the south-west of WA.

Introduced fish are a significant problem in places. Gambusia (*Gambusia holbrooki*), introduced from the Americas, eats a wide variety of native species. In eastern Australia it has detrimentally affected the threatened green and gold bell-frog (*Litoria aurea*), but in WA its effects have not been researched. The introduction of redfin perch (*Perca fluviatilis*) to many WA waterways is affecting native fish. In Tasmania, the introduced brown trout (*Salmo trutta*) and rainbow trout (*Oncorhynchus mykiss*) have eliminated native fish from many rivers. In WA, probably because the highly temporary nature of our rivers makes them less-than-ideal habitat for trout, this introduced predator does not seem to have had such a dramatic effect, however, there is still cause for concern.

Another invasive species that is on its way to WA is the cane toad (*Bufo marinus*), introduced into Queensland in 1935. It will establish throughout the wetter parts of the Kimberley, where it can be expected to greatly affect some native animals via its highly toxic skin glands (see Chapter 7).

Invasive invertebrate predators include a species of snail from Europe, *Oxychilus* sp., which has become established in the south-west, where it is eliminating native snail species. Many species of native snails may be at risk from this predator. Predation by *Oxychilus* is thought to be at least partly responsible for the extinction of the Albany snail (*Helicarion castanea*) and the Pemberton snail (*Occirhenea georgiana*) (see Chapter 9). Several species of slugs have also been introduced, some of which may be predators. Introduced ants, such as the Argentine ant (*Linepithema humilis*), Singapore ant (*Monomorium destructor*) and coastal brown ant (*Pheidole megacephala*), have eliminated native ants from much of metropolitan Perth and are now turning up in bushland areas. Introduced earthworms are probably eliminating native earthworms. European wasps (*Vespula germanica*), already found occasionally in Perth, and American fire ants (*Solenopsis invicta*), which have established in suburban Brisbane, have the potential to do great damage to WA's native wildlife (and to our way of life) should they become widespread.

Herbivores

Many grazing animals have established in Australia. Some were brought in as stock, others were released for food or sport, and still others arrived on ships or in food and equipment. The list seems endless: sheep, goats, cattle (both European cattle and zebu), horses, donkeys, camels, rabbits, pigs and water buffalo. Even black buck (*Antilope cervicapra*) and red deer (*Cervus elephus*) were released in WA, but fortunately failed to establish. Introduced herbivores compete with native species for food, and by damaging plants degrade habitat as well. Rabbits have been the most destructive of the introduced herbivores, eliminating palatable plants, and the animals that depend on them, from vast areas of Australia. The effect of introduced herbivores is not restricted to farming and pastoral areas, as species such as rabbits, goats, pigs, donkeys, cattle and camels are widespread in unoccupied lands, including conservation reserves.

The honey bee (*Apis mellifera*), another introduced species, competes with native insects and vertebrates for pollen and nectar, and may also reduce the pollination rate of some specialised native plants. By establishing hives in hollows in trees and elsewhere, they reduce the availability of nesting hollows to native birds (such as cockatoos and parrots) and mammals (such as possums). Rainbow lorikeets (*Trichoglossus haematodus*)—now well established in Perth with the potential to spread further—also compete with native birds for nesting hollows. Eastern long-billed corellas (*Cacatua tenuirostris*), now established in Perth suburbs from escaped cage birds, could, if they moved outside the metropolitan area, hybridise with the threatened Muir's corella (*Cacatua pastinator pastinator*) (see Chapter 5), and their eradication is urgent.

Diseases

Diseases introduced to Australia can affect native species. Diseases of wild animals in Australia are not well understood, but there is significant concern. Several accounts of large numbers of marsupials dying, apparently from disease—in places like the Nullarbor Plain and the south-west—during the first half of the twentieth century, suggest that introduced diseases may have played a role in the decline of native mammals, but there are no scientific studies to support this.

Toxoplasma gondii, a protozoan parasite that causes toxoplasmosis, is transmitted by cats and can cause blindness, birth defects, miscarriage and possibly mental disease in people. It leads to blindness and damage to the central nervous system and respiratory organs of native wildlife. Some native animals, such as bandicoots and kangaroos, never recover from the disease.

The origin of the recently-discovered chytrid fungus (*Batrachochytrium dendrobatidis*) is not known, but it was probably introduced from Africa or South America. It causes a disease known as chytridiomycosis in frogs, and is thought to have caused the extinction of some

Above The eastern long-billed corella has established in Perth from aviary escapes and presents a threat to the Endangered Muir's corella.

Photo – Babs and Bert Wells /CALM

frog species in eastern Australia. The fungus is present in many species of frogs in the south-west of WA, but so far does not seem to be causing significant decline in these species.

The recent discovery of bat lyssavirus in Australia shows how little we know about diseases in wildlife. This virus, closely related to rabies, is present in many species of bats, particularly in the tropics, and has caused the death of at least one person. People should avoid handling bats (both flying-foxes and small insectivorous bats), especially in the tropics, as the disease can be transmitted by biting.

Elsewhere in the world, it has been shown that indigenous animal diseases become more common when ecosystems and species are put under stress, such as in fragmented landscapes. There are, as yet, no studies on this issue in Australia.

One major disease of native plants, dieback disease caused by the root-rot pathogen *Phytophthora cinnamomi*, affects native animals by altering habitats. In the Stirling Range National Park and along the south coast, species-rich kwongan (heath) dominated by proteaceous genera like *Banksia*, *Dryandra* and *Grevillea* is being converted to a species-poor sedgeland. The many mammals, birds and insects that depend on the susceptible native plant species are thus eliminated. Recent research into ngilkait (Gilbert's potoroo) at Two Peoples Bay Nature Reserve shows that it survives only in areas where *Phytophthora* is absent. This may be because the fungi on which it feeds are eliminated by the pathogen and/or because its shrubland habitat changes to a sedgeland.

Weeds

Hundreds of alien plants have become established in bushland areas in WA—in 1999 there were 1350 species of potential and existing environmental weeds in the WA Herbarium's database. Environmental weeds can greatly alter ecosystems, thus degrading and altering animal habitat. Weed invasion may also lead to changes in fire regimes, prevent native plants from germinating and even smother native plants to the extent that few of them remain.

With such a variety of climates and ecosystems, weeds in WA tend to have localised effects, but some species have invaded large areas. Buffel grass (*Cenchrus ciliaris*) is one such weed. Introduced from Africa as a food plant for stock, it has invaded many, especially coastal, ecosystems and islands in the north-west of WA, eliminating native plants and changing fire regimes. Bridal creeper (*Asparagus asparagoides*) is a widespread, suffocating weed of the south-west, and Victorian tea-tree (*Leptospermum laevigatum*) is another major bushland weed that forms dense monocultures under which most native species cannot exist. On the Swan Coastal Plain near Perth, introduced grasses such as perennial veldt grass (*Ehrharta calycina*), wild oat (*Avena barbarta* and *A. fatua*) and other introduced grasses are changing the understorey of many remnant bushland areas and must be affecting the area's original fauna.

Below Weeds being eradicated from Canning River Regional Park. Weeds choke out natural vegetation and are a major threat to biodiversity in Western Australia.

Photo – Michael James/CALM

Climate change

It is now widely accepted that the world's climates are changing due to the emission of 'greenhouse gases' such as carbon dioxide and methane. The latest information on climate change developed by the Intergovernmental Panel on Climate Change, established by the World Meteorological Organization (WMO) and the United Nations Environment Program (UNEP), can be found at www.ipcc.ch. Predictions by CSIRO suggest that the climate of the south-west of WA—one of the world's 25 megadiverse biodiversity hotspots—will become drier and hotter. This could have major impacts on our native plants and animals. Along the south coast of WA, for example, there are a series of peaks up to 1000 m above sea level. One hundred and one species of flowering plants, several ecological communities and a significant number of invertebrates that are relicts from Gondwanan times are restricted to one or a few mountaintops. Predictions suggest that, relative to 1990, south coast temperatures will rise by up to 1.5°C and rainfall will decrease by up to 20% by 2030. By 2070, temperatures may rise by up to 5°C and rainfall may reduce by up to 60%. As there is nowhere for the mountaintop endemics to go, it is likely that many species will become extinct.

Lower rainfall will result in less run off into rivers and wetlands, and lower groundwater levels. Wetlands will be greatly reduced and many will become dry. Aquatic organisms, including stygofauna (see Chapter 9) and some frog species, may be affected. Marine ecosystems will not be excluded. Many coral reefs will disappear or become 'deserts', due to coral bleaching caused by higher sea temperatures, and minor changes in ocean currents could have a dramatic effect on fish numbers and the seabirds and marine mammals that depend on them.

Climate change, especially in the south-west, will alter fire regimes by extending the period when vegetation is dry enough to burn. Reduced rainfall and warmer temperatures will also affect the recovery time of ecosystems and organisms following disturbances such as fire. Climate change could also result in outbreaks of pests and diseases, and there is some evidence that decline of trees, such as wandoo and tuart, in the south-west is linked to water stress.

Other threatening processes

Over-exploitation (through hunting or fishing) does not seem to have been a direct cause of extinctions in WA so far. However, two species of shark—the great white (*Carcharodon carcharias*) and grey nurse (*Carcharias taurus*)—are listed because of this threat. Over-collecting of attractive animals, such as leafy seadragons and large, colourful shells of marine molluscs, and illegal collecting and smuggling of native parrots and cockatoos are issues that require attention. Over-harvesting of several species of marine turtles has been the primary cause of them now being listed as threatened worldwide. While over-harvesting of turtles is not currently thought to be a threat in WA, most turtles that nest in the State forage over exceptionally long distances, often in places away from our shores where they may be hunted unsustainably.

'Bycatch' associated with commercial fisheries is an increasing problem. The drowning of albatrosses by the longline fishing industry is of enormous concern, and Australia has

Above Stirling Range National Park. Climate change could lead to the extinction of invertebrates restricted to mountaintops.

Photo – Jiri Lochman/CALM

been at the forefront of international efforts to reduce the impacts of fishing on seabirds. While some progress has been made, many fishing boats registered in nations not interested in conservation still kill thousands of seabirds every year. The 13 albatross species and two petrel species listed as threatened in WA are all threatened by longline fishing. Longline fishing is also affecting marine turtles. Bycatch by trawling, especially for prawns, has received recent attention, and most otter trawls now include a turtle exclusion device (TED), also known as a bycatch reduction device (BRD). Net fishing is threatening species of freshwater sawfish and spear-tooth sharks in northern Australia, as well as marine and estuarine sawfish and whiprays.

Timber harvesting in native forests has been a controversial activity for some time. Most studies in WA suggest it has little or no long-term effect on threatened species, but recent research suggests that the ngwayir (western ringtail possum) is severely affected by timber harvesting. Some species that depend on old trees for nesting sites—especially Baudin's black-cockatoo and the forest red-tailed black-cockatoo—may be under threat from logging, and this needs further study. Invertebrate animals that depend on sites that may be disturbed by timber harvesting activities also need further research. The tingle trapdoor spider, already listed as threatened, is one such highly-restricted species, but, fortunately, it is found in a national park. Other restricted, specialised species that occur in State forests may not be in the same situation.

Human over-utilisation of any resource on which animals also depend can lead to species becoming threatened. As well as some of the issues discussed above, over-utilisation can include

Below Australia has been at the forefront of efforts to reduce the impact of fishing on seabirds such as albatrosses, which take baited hooks and drown.

Photo – Rosemary Gales/ Tasmanian Nature Conservation Branch

pumping of groundwater for mine dewatering or human use, wildflower harvesting, mining and so on. Lowering of groundwater levels in the Gnangara Mound on the northern edge of Perth, through planting of introduced pines and pumping for human consumption, is threatening unique animal communities in Yanchep National Park (see p. 170). Damming of rivers prevents some migratory fish from moving upstream to breed—in the south-west of WA damming has significantly affected the pouched lamprey (*Geotria australis*) and the mud minnow (*Galaxiella munda*).

Interaction of threatening processes

Often, threatening processes combine to produce a much greater effect than would otherwise be expected.

Weeds can lead to altered fire regimes. More frequent fire opens up habitat and allows invasion by feral animals. Extensive fire can restrict animal populations to small areas, making them more susceptible to predation by foxes and cats, and so on. Thus, threatening processes cannot be managed in isolation—a holistic approach is needed.

We need to live in our environment in a manner that does not threaten other species. Sustainable development principles are starting to be incorporated more into our economic activities, but are proving difficult for our society to achieve.

The next chapter examines what is being done and what should be done to combat threatening processes.

Above It is likely that feral cats have been responsible for the extinction of several species of native animals and they continue to wreak havoc on our native wildlife.

Photo – Babs and Bert Wells /CALM

References

Abbott, I. and Burrows, N. (eds) (2003). *Fire in ecosystems of south-west Western Australia: impacts and management.* Backhuys Publishers, Leiden, The Netherlands.

Burbidge, A.A. and Manly, B.F.J. (2002). Mammal extinctions on Australian islands: causes and conservation implications. *Journal of Biogeography* 29, 465–474.

Burrows, N.D., Burbidge, A.A. and Fuller, P.J. (in press). Nyaruninpa: Pintupi burning in the Australian western desert. Proceedings of international symposium on native solutions: indigenous knowledge and today's fire management. Hobart, Tasmania, July 2000.

Dickman, C.R. (1996) *Overview of the impact of feral cats on Australian native fauna.* Australian Nature Conservation Agency, Canberra.

Kinnear, J.E., Onus, M.L. and Bromilow, R.N. (1988) Fox control and rock-wallaby population dynamics. *Australian Wildlife Research*, 15, 435–477.

Kinnear, J.E., Onus, M.L. and Sumner, N.R. (1998) Fox control and rock-wallaby population dynamics – II. An update. *Wildlife Research*, 25, 81–88.

Latz, P. (1995). *Bushfires and bushtucker: Aboriginal plant use in central Australia.* IAD Press, Alice Springs.

Veitch, C.R. and Clout, M.N. (eds) (2002). *Turning the tide: the eradication of invasive species.* IUCN SSC Invasive Species Specialist Group, Auckland.

Conserving threatened fauna in Western Australia

What is being done to prevent extinctions in Western Australia and bring species back to levels of abundance at which they will no longer need special conservation programs?

Legislation

Modern biodiversity conservation legislation, which includes provisions for conserving threatened species and ecological communities, is an important basis for developing conservation programs. For a long time, the *Wildlife Conservation Act 1950* (as amended) has provided the legal basis for biodiversity conservation programs in WA, but it is now outdated. While it provides special protection for threatened animals, it does not protect their habitat or include provisions relating to recovery plans. Ecological community conservation, a relatively new concept, is not mentioned. Hence, a Biodiversity Conservation Bill—aimed at replacing the Wildlife Conservation Act—is now being prepared for introduction to Parliament.

Conservation through reservation

As land clearing has been the most significant threatening process, the creation of a comprehensive, adequate and representative (CAR—see p. 26) conservation reserves system is the most important means of preventing extinctions.

The creation of conservation reserves—national parks, nature reserves, conservation parks, marine parks and equivalent areas—has long been a primary conservation technique. At 30 June 2003, WA had a terrestrial conservation reserve system comprising 5,095,378 ha in 69 national parks, 704,216 ha in conservation parks, 10,827,256 ha in nature reserves and 230,613 ha in miscellaneous conservation reserves under the *Conservation and Land Management Act 1984*. The total area of 16,857,463 ha equates to 6.7% of WA's land area. In addition, 4,777,161 ha of former pastoral land had been acquired for conservation reserves as at 30 June 2003, but were not yet reserved, and a further 698,000 ha (mostly State forest) will be formally reserved in accordance with the forest management plan that implements the Government's 'Protecting our old-growth forests policy'. Reservation of these areas will bring the total conservation reserve system to more than 21,000,000 ha, or about 8.5% of WA's land area. This figure does not include State forest, timber reserves and some miscellaneous lands also managed by the Department of Conservation and Land Management. The largest of these categories is State forest, with an area of 1,729,974 ha.

Because of historical land use decisions, the reserve system does not meet the CAR criteria. The national framework against which progress towards a CAR reserve system is planned and measured in the terrestrial environment is the Interim Biogeographic Regionalisation of Australia (IBRA). Under IBRA, WA has been divided into 26 bioregions (of which eight are shared with the Northern Territory and South Australia) and 52 sub-regions or provinces. Of the 26 bioregions in WA, eight are ranked nationally as very high or high priority for further reservation to meet major gaps at a national scale in the reserve system, 11 as moderate priority and seven as low priority. However, even regions ranked as 'low' contain additional areas warranting reservation to protect special values (such as threatened species and ecological communities), as well as areas of amenity or scenic value and community attachment. Some sub-regions have no conservation reserves at all.

Above A fox-proof fence constructed at Ellen Brook Nature Reserve to protect the western swamp tortoise.

Opposite University of Western Australia western swamp tortoise researcher Gerald Kuchling downloading data on temperatures within a western swamp tortoise underground nest.

Photos – Andrew Burbidge /CALM

Attributes of a comprehensive, adequate and representative conservation reserve system

The terms 'comprehensive', 'adequate' and 'representative' (CAR) together describe the attributes of an ideal reserve system. These terms were defined in the Australian and New Zealand Environment and Conservation Council 1999 Guidelines for Establishing the National Reserve System as:

- comprehensiveness – inclusion of the full range of ecosystems recognised at an appropriate scale within and across each bioregion;

- adequacy – the maintenance of the ecological viability and integrity of populations, species and communities; and

- representativeness – the principle that those areas that are selected for inclusion in reserves reasonably reflect the biotic diversity of the ecosystems from which they derive.

In addition to using the scientifically-based CAR criteria, spectacular landforms and scenery, as well as natural areas of high public use, are also commonly included in parks and reserves.

WA still has a long way to go to meet the CAR criteria (see www.naturebase.net/projects/pdf_files/car _report.pdf), and various strategies are underway to improve the system (see Chapter 11).

Below Bushwalkers in John Forrest National Park.

Photo – Gordon Roberts /CALM

Table 7 summarises information on the proportion of land reserved for nature conservation in all bioregions in WA, as well as providing information on the number of vegetation types remaining in each bioregion and the number in reserves. In this table 'reserves' includes land in IUCN categories I–VI, so the figures include State forest. Definitions of IUCN categories can be found at http://www.unep-wcmc.org/protected_areas/categories/eng /index.html and information on IBRA and the national reserves system at http://www.deh.gov.au/parks/nrs/.

Habitat protection through reservation is also important in the marine environment. The declaration of marine conservation reserves did not get underway in any significant manner in WA until the 1990s, after Parliament passed amendments to the Conservation and Land Management Act. Since then, the Department has worked to develop a system of marine conservation reserves, guided by the 1994 report of the Marine Parks and Reserves Selection Working Group.

At 30 June 2004, there were seven marine parks with an area of 1,096,316 ha and one marine nature reserve with an area of 132,000 ha. Table 8 provides information on the areas in Interim Marine and Coastal Regionalisation of Australia (IMCRA) regions in WA's coastal waters currently in marine protected areas and projects what may happen if areas currently being considered for reservation are declared. To date, 9.7% of WA coastal waters are reserved. Most areas within marine conservation reserves are zoned for multiple use with a relatively small proportion in sanctuary zones, where fishing and collecting are totally prohibited.

Above The Bibbulmun Track near Point Irwin in the Warren bioregion.

Below Morgan Falls in Drysdale River National Park, northern Kimberley bioregion.

Photos – Andrew Burbidge /CALM

Table 7   Area of land reserved in IUCN Categories I–VI in IBRA bioregions occurring in WA.

Bioregion	Area in bioregion (ha)	Area reserved* (ha)	Per cent of bioregion reserved*
Avon Wheatbelt	11,167,095	124,379	1.1%
Carnarvon	9,215,451	695,206	7.5%
Central Kimberley	8,081,153	357,951	4.4%
Central Ranges	5,221,498	0	0.0%
Coolgardie	14,332,216	1,339,065	9.3%
Dampierland	8,813,000	91,239	1.0%
Esperance Plains	3,507,281	966,278	27.6%
Gascoyne	19,988,095	2,076,344	10.4%
Geraldton Sandplains	4,534,213	581,113	12.8%
Gibson Desert	17,236,797	2,067,479	12.0%
Great Sandy Desert	31,809,961	854,914	2.7%
Great Victoria Desert	24,780,891	2,325,837	9.4%
Hampton	1,229,072	134,486	10.9%
Jarrah Forest	5,416,008	316,036	5.8%
Little Sandy Desert	12,192,776	558,448	4.6%
Mallee	8,853,043	1,493,453	16.9%
Murchison	31,920,977	655,324	2.1%
Northern Kimberley	8,620,134	1,085,329	12.6%
Nullarbor	16,021,285	2,584,062	16.1%
Ord Victoria Plain	5,823,015	342,969	5.9%
Pilbara	19,310,548	1,614,407	8.4%
Swan Coastal Plain	1,858,978	159,885	8.6%
Tanami	3,214,599	0	0.0%
Victoria Bonaparte	1,932,467	107,978	5.6%
Warren	1,027,586	296,864	28.9%
Yalgoo	4,858,847	492,838	10.1%

* includes IUCN I–IV and land purchased for nature conservation but not yet reserved

data correct at September 2001

Table 8 Existing and projected CALM Act marine conservation reserves by IMCRA region.

IMCRA meso-scale bioregion	Area of meso-scale bioregion within WA coastal waters (ha)	Gazetted area currently in reserve (ha)	% of meso-scale bioregion in reserves	Government priority area in progress for reservation (ha)	Projected % of bioregion in reserves when areas are added *
Cambridge-Bonaparte	500,942	0	0.00%	0	0.00%
Bonaparte Gulf	167,364	0	0.00%	0	0.00%
Kimberley	3,246,602	0	0.00%	0	0.00%
King Sound	595,923	0	0.00%	0	0.00%
Canning	414,690	0	0.00%	0	0.00%
Eighty Mile Beach	168,819	0	0.00%	0	0.00%
North West Shelf	22,075	0	0.00%	0	0.00%
Oceanic Shoals	164,496	23,250	14.13%	64,129	53.12%
Pilbara (nearshore)	1,568,191	6835	0.44%	214,083	14.09%
Pilbara (offshore)	1,099,534	20,984	1.91%	240,957	23.82%
Ningaloo	215,527	197,745	91.75%	17,782	100.00%
Zuytdorp	381,136	140	0.04%	20,218	5.34%
Shark Bay	1,424,959	880,595	61.80%	0	61.80%
Abrolhos Islands	242,664	0	0.00%	0	0.00%
Central West Coast	383,538	91,876	23.95%	0	23.95%
Leeuwin-Naturaliste	368,649	6891	1.87%	122,786	35.18%
WA South Coast	1,291,550	0	0.00%	0	0.00%
Eucla	376,640	0	0.00%	0	0.00%
TOTAL	**12,633,297**	**1,228,316**	**9.7%**	**760 959**	**15.75%**

* Projected % of bioregion in reserves when areas currently in progress for reservation are added. Data correct at August 2003.

Below The Montebello Islands. A proposal for two new marine parks and the State's first marine management area in the Montebello Islands–Barrow Island area is under consideration.

Photo – Kevin Kenneally /CALM

Conservation outside reserves

Even an ideal reserves system will not include habitat for all species, especially rare ones. Many threatened animals depend on habitat on private land–Carnaby's black-cockatoo is a good example (see p. 95). Biodiversity conservation must also be carried out outside reserves to be effective. Increasingly, private landowners are taking responsibility for native plants and animals on their land. The Department of Conservation and Land Management has developed programs to help farmers and other landholders conserve local species. 'Land for Wildlife' is one such program. This voluntary program recognises the conservation efforts of private landholders

Above CALM scientist Adrian Wayne releasing a threatened western ringtail possum in the forest near Manjimup.

Photo – Adrian Priest

and managers, and helps them do more to conserve native plants and animals on their lands by protecting, expanding or creating suitable habitat. Landholders can register land under the scheme. At April 2003, 1159 properties were registered with Land for Wildlife covering an area of 662,862 ha, of which 235,110 ha was remnant vegetation. Covenanting schemes—which allow landowners to enter into agreements that will protect parts of their land after they have sold it—are also becoming increasingly popular. At the end of April 2003, the Department's covenant program had 27 registered covenants protecting 635 ha. Forty-four further covenants were being negotiated covering 7813 ha.

Privately-owned conservation reserves are starting to appear, with organisations like Australian Wildlife Conservancy and the Australian Bush Heritage Fund purchasing land to ensure its long-term protection for biodiversity conservation.

The recovery process

The 'recovery process' is a set of steps used to guide the identification, prioritisation and conservation of threatened species. It involves:

1. reviewing the conservation status of species or subspecies;

2. preparing priority lists of threatened species or subspecies;

3. conducting the necessary research; and

4. producing costed recovery plans.

Then, for each recovery plan:

5. obtaining funding;

6. implementing the plan; and

7. monitoring and reviewing the plan.

Not all steps may be necessary for all species (for example, enough may already be known about the biology and ecology of the species to draft a recovery plan without doing further scientific research), but the process is a useful framework to guide conservation of threatened species. Often, when there is insufficient information about a highly-threatened species or ecological community, the Department of Conservation and Land Management prepares an 'interim recovery plan' that prescribes the necessary steps to minimise the risk of extinction in the short term, and details the research needed to develop a full recovery plan.

Recovery plans should:

- provide a brief background about the species (or ecological community), including a summary of relevant aspects of its biology and ecology, and threats to the species;

- state an objective to be achieved within a given timeframe;

- provide quantifiable/measurable criteria for success and failure;

- prescribe actions needed to ensure the conservation of the species;

- estimate total costs over a selected period—usually five or ten years;

Above University postgraduate student Harriet Mills in the dibbler breeding precinct at the Perth Zoo. Captive breeding at the zoo has greatly assisted in the recovery of several threatened species.

Photo – Marie Lochman

- if possible, identify sources of funding;

- allocate responsibility for implementing the actions; and

- ensure that the species is monitored and that the plan is modified if required in the light of implemented actions and new information.

In the descriptions of each species and ecological community in the subsequent chapters, I have stated whether the conservation of a species or community is guided by a recovery plan.

Before a recovery plan can be developed and implemented, however, a 'recovery team' should first be appointed. As a general rule, recovery teams should include representatives of all groups that have a stake in or may be affected by the recovery plan. These include, where appropriate:

- Departmental staff involved in managing or researching the species or ecological community;

- staff from other State Government agencies involved in the recovery of the species, such as Perth Zoo, Kings Park and Botanic Garden, Main Roads, or the WA Water Corporation;

- external scientists with expertise in the species or ecological community, for example, from universities or CSIRO;

- representatives from local Government;

- landowners or land managers, preferably representing a local organisation such as a Land Conservation District Committee or catchment management group;

- representatives of funding agencies or sponsors; and

- members of non-Government conservation groups involved in conserving the species, such as local naturalists clubs.

Members are appointed in their own right, and are not appointed by any organisation of which they happen to be members. Recovery teams should not include people who are opposed to the conservation of the species or community— there are other forums where such views can be expressed via the political process. Most recovery teams meet twice per year.

Because there are many threatened species and threatened ecological communities, arranging for a team to look after multiple species is efficient where appropriate. Thus, one recovery team coordinates conservation work on the two *Geocrinia* frogs from near Margaret River, and one team manages five species of threatened birds.

Most recovery plans and teams have been extremely effective in recovering threatened species. Some of the details are discussed in later chapters.

Combating threatening processes

Recovery plans include specific conservation actions needed for the recovery of the species concerned. The actions required will differ from species to species, but commonly include habitat protection and management, introduced predator control, population monitoring and additional research aimed at developing new and/or more cost-effective conservation actions. In some cases, captive breeding and/or translocation may be required to rapidly build up the numbers of a species in the wild.

Sometimes it will be more effective to combat broadscale threatening processes than to try to deal with the same threat on a species-by-species basis. Introduced predators are one threat that can be addressed in such a manner through large parts of the State.

Western Shield

Pioneering research by Department of Conservation and Land Management scientists in the 1970s and early 1980s showed that foxes were having an enormous impact on some species of native mammals, and that fox control led to a resurgence in numbers of remnant mammal populations. Later, it was shown that locally extinct species could be reintroduced once foxes had been controlled. This led to the introduction, in 1995, of *Western Shield*. In 2003, more than 3,000,000 ha of conservation lands were being baited for foxes, using dried meat baits with '1080'. When it is used carefully, 1080 is a benign toxin in WA, especially in the south-west, where it occurs naturally in poison peas (*Gastrolobium* species), and native animals have thus evolved a high degree of resistance to it.

However, when predator control was attempted in arid parts of the State it was discovered that, in the absence of foxes and dingoes, feral cats became much more common and prevented native species from recovering. While cat control was mooted in *Western Shield* from the start, this discovery led to increased research into methods of broadscale feral cat control. This work is ongoing and includes research in a number of different environments around the State, including the central deserts.

While much of this research has been successful, and is leading to the availability of effective baits to control feral cats, further work is needed before the bait can be registered for broadscale use. Cats have already been eliminated from several islands, and research has shown that cat numbers can be greatly reduced over large areas of arid land.

Other threats

Other broadscale threats that are being researched and managed include secondary salinisation in the south-west, dieback disease caused by *Phytophthora* and inappropriate fire regimes. Ridding island nature reserves of introduced animals has been another theme. Six exotic mammals—the red fox, feral cat, goat, rabbit, black rat and house mouse—have now been eradicated from more than 45 islands in a series of projects since the 1960s. Most effort has been directed at black rats, with more than 31 islands now clear of this species and a major project in the Montebello Islands hopefully nearing conclusion.

Top CALM researchers preparing feral cat baits in the Gibson Desert.

Above Feral cat.

Photos – Neil Burrows/CALM

References

Kinnear, J.E., Onus, M.L. and Bromilow, R.N. (1988). Fox control and rock-wallaby population dynamics. *Australian Wildlife Research*, 15, 435–477.

Kinnear, J.E., Onus, M.L. and Sumner, N.R. (1998). Fox control and rock-wallaby population dynamics – II. An update. *Wildlife Research*, 25, 81–88.

Veitch, C.R. and Clout, M.N. (eds). (2002). *Turning the tide: the eradication of invasive species.* Proceedings of the International Conference on Eradication of Island Invasives. IUCN SSC Invasive Species Specialist Group. IUCN, Auckland.

Wilson, B.R. (1994). *A representative marine reserve system for Western Australia: report of the Marine Parks and Reserves Selection Working Group.* Department of Conservation and Land Management, Como.

Mammals

Mammals account for the largest number of known extinctions in any group of Australian animals since European settlement in 1778. Twenty-two species are extinct, with another two (possibly three) species disappearing from Christmas Island, an Australian External Territory. Eleven of the extinct species lived in Western Australia. Six additional subspecies of mammals are extinct in Australia, five of which occurred in WA. Without islands, this conservation disaster would have been even more serious, as eight once-widespread species disappeared from mainland Australia (six of them occurred in WA) and became restricted to islands. Some of these have been, or are currently being, reintroduced to mainland sites.

Above Bilbies are one of very few medium-sized mammals that have not become extinct in arid areas.

Photo – Michael James/CALM

Opposite Kingo, or red-tailed phascogale, formerly occurred through much of arid and semi-arid southern Australia, but are now confined to a few areas of remnant vegetation in the WA Wheatbelt.

Photo – Babs and Bert Wells /CALM

A high proportion of WA's native mammals is listed as threatened with extinction. Of a total of 194 native mammal species recorded in the State, 34 (17.5%) are listed as threatened, in addition to the 11 (5.7%) that are extinct. If marine mammals are excluded, 29 (18.5%) of 157 species are listed, in addition to the 11 (7.0%) extinct species. Thus, more than one quarter of WA's terrestrial mammal species are extinct or threatened with extinction. Of the 37 marine species, four (10.8%) are threatened, as is the only additional subspecies. As well as the 194 species of native mammals in WA, another 16 subspecies are widely recognised by mammal experts as valid, a total of 210 taxa (species or subspecies). Eight subspecies are listed as threatened, and another four are extinct. The proportion of species and subspecies listed as extinct is 7.1% and the proportion listed as threatened is 20.0%, a total of 27.1%.

In a 1989 scientific paper, Norm McKenzie and I drew attention to the correlation between the number of extinct and threatened Western Australian mammals and rainfall—the more arid regions had higher rates of extinction and decline—and also to a correlation with body weight. We coined the term 'critical weight range' (CWR) for non-flying species with mean adult body weights of between 35 g and 5500 g, and noted that all but two of Australia's extinct mammals and most threatened mammals were within the CWR.

A more recent analysis of Australia's 305 species of native terrestrial mammals was undertaken as part of a National Biodiversity Audit. We examined the contribution of rainfall, body weight, ability to fly, and habitat destruction and degradation to the status of mammal species in the 85 Interim Bioregionalisation of Australia regions. We found that rainfall explained 48% of the accounted-for pattern of mammal decline, body weight explained 19% and ability to fly 18%. Habitat destruction and degradation explained only 15%. The last result wasn't surprising, as many of Australia's extinct and threatened mammals have disappeared from national parks and uncleared bushland, as well as from agricultural and pastoral land. So, while habitat destruction is an important threatening process for mammals, other threats have had a much greater effect.

The correlation between mammal decline and body weight is thought to be due to the limited mobility and relatively high metabolic requirements of CWR species, with environmental changes since settlement (Chapter 2 describes the diversion of environmental resources to people and introduced herbivores, the reduction in vegetative cover by introduced herbivores and changed fire regimes) emulating increased aridity. Mammals in the CWR seem particularly susceptible to predation by foxes and feral cats, with these introduced species being more successful hunters in arid, more open, environments. These conclusions, if correct,

Above The Barrow Island euro occurs on a single island.

Photo – Jiri Lochman

Below Rock-wallabies have become threatened because of predation by introduced foxes. This juvenile has lost part of its tail to a fox.

Photo – Babs and Bert Wells /CALM

have serious ramifications for increased mammal extinction rates in areas of the country where climate change is predicted to bring about reduced rainfall.

Most of the extinct mammals disappeared many decades ago, so we know little about them. Fortunately, collectors working for European museums made significant collections, especially in the south-west. John Gilbert, for example, collected for John Gould between 1838 and 1845, and Guy Shortridge collected for the British Museum between 1904 and 1907. Apart from notable exceptions, such as Otto Lipfert's efforts along the Canning Stock Route in the 1930s and Hedley Finlayson's extensive searches in central Australia, also in the 1930s, mammals in the sandy deserts were barely surveyed until four-wheel-drive vehicles became readily available and access tracks were built in the 1960s. There was little systematic scientific survey for mammals in WA until very recently. Only in the second half of the twentieth century did a government

conservation department begin to research and help to conserve mammals, and the WA Museum had few scientific staff until the 1960s. Once comprehensive biological surveys began in the late 1960s and 1970s, much new information was discovered about our native mammals, and new species were found.

Two techniques—subfossil research and oral history—have helped to overcome, to some extent, the lack of historical information. Subfossils are skeletal remains, retrieved from cave floors, sinkholes and other areas, of animals that have died in the past few hundred years. The bones are not fossilised in the sense that they have not changed chemically. Subfossils may accumulate in caves, into which animals have fallen and died, or may be derived from predator dens and roosts, especially those of owls. Alex Baynes, Research Associate of the WA Museum, sometimes with support from the Department of Conservation and Land Management, has developed considerable expertise in finding and

identifying subfossil mammals in WA, and has greatly extended our knowledge of the former distribution of many uncommon and extinct mammals.

Oral history research has been particularly rewarding in the western deserts, involving Aboriginal people who had previously lived a 'hunter gatherer' lifestyle and depended on desert animals and plants for food. This technique was pioneered in Australian deserts in the 1930s, by exceptional South Australian mammal expert Hedley Finlayson, and applied again when Phil Fuller and I started oral history research at Warburton in 1976. In the early 1980s, we extended our work to as many desert Aboriginal communities as possible, expanding our studies to the Northern Territory and South Australia with colleagues Ken Johnson and Ric Southgate from the Northern Territory conservation agency.

Historical information from early settlers has also been valuable. Many settlers and explorers kept diaries, some of which have been published. For example, Bruce Leake's *Eastern wheatbelt wildlife* provides detailed information on mammals in the Kellerberrin area in the late 1800s and early 1900s.

These techniques, along with more recent mammal surveys, have provided a fairly good understanding of the conservation status of most native mammals. However, there is still much to learn, especially in relation to biology, ecology and the development of cost-effective conservation techniques. We still know very little about several species, such as the northern phascogale, kakarratul and itjaritjari (both marsupial-moles), djintamoonga (black-footed tree-rat), dayang (heath mouse) and Butler's dunnart. Amazingly for a developed country, new species of mammals are still being discovered and described in WA, and further research is required to clarify the taxonomic status of several species.

Mammals of the central deserts have declined greatly over the past 100 years or so. About a third of the native mammals of the western desert have disappeared completely, while others are highly threatened. Studying and conserving threatened desert mammals is difficult and time consuming, as populations fluctuate considerably in abundance, depending on rainfall and other factors. Some studies on species such as mulgaras and bilbies began when local populations were reasonably common, only to find that the animals disappeared for no apparent reason. The two species of marsupial-mole are particularly difficult to study due to their cryptic underground habits (they live up to 2 m below the surface) and apparent scarcity.

Common names are a frequent cause of argument among naturalists. International rules apply for scientific names, but common names are just that— the names used commonly by local people. Where appropriate, this book uses Western Australian Aboriginal names as alternatives to 'English' names, to try to promote 'Australian' names for Australian animals. I have also provided alternative names, including Aboriginal names where these are known. Pronunciation rules are provided in the books and articles in the references at the end of the chapter. Aboriginal words are fairly easy to pronounce once the basic rules are understood.

Top Mulgaras inhabit burrows to escape the desert heat, emerging at night to feed.

Photo – Babs and Bert Wells /CALM

Above Still abundant on Rottnest and Bald islands, quokkas have declined on the mainland.

Photo – Ken Stepnell/CALM

Extinct species and subspecies

Marl, western barred bandicoot (mainland subspecies)
Perameles bougainville fasciata

Other names: Mal/marl, nymal
(Nyoongar).

Description: This extinct subspecies was
similar to the western barred bandicoot
from the Shark Bay islands (see p. 59),
but had stronger colouring and body
stripes. Marl occurred from Onslow
southwards to near Perth, through the
Wheatbelt, across the Nullarbor Plain and
through South Australia to western New
South Wales and western Victoria.

The subspecific taxonomy within
P. bougainville is unclear, and the
populations from the Shark Bay islands
may be the same subspecies as those
from mainland south-west WA (in which
case all WA animals would be
P. bougainville bougainville), while the
eastern populations (now extinct) may be
P. bougainville fasciata.

Approximate date of extinction:
Marl were last collected on the mainland
in 1906, in the south-west of WA. Oral
history information has suggested that
they may have survived on the Nullarbor
Plain until about 1950.

Probable cause of extinction: Feral
cats and foxes, together with competition
with and habitat destruction by sheep
and cattle.

Below Western barred
bandicoots disappeared from
the mainland by the middle
of the twentieth century.

Illustration – From *The
Mammals of Australia* by
John Gould (1863)

Kantjilpa, pig-footed bandicoot *Chaeropus ecaudatus*

Other names: Boda, woda, boodal (Nyoongar), kalatawurru, kanytjilpa, marakutju, parrtiriya, takanpa, tjunpi, yirratji (western desert Aboriginal dialects).

Description: This small bandicoot had long, slender limbs with two functional toes on each front foot and one on each rear foot. The head and body length was 230–500 mm and the tail length 100–140 mm. The hair, though coarse, was not spiny. The upper parts were grizzled grey, tinged with fawn, or almost orange brown, and the underparts were white or pale fawn. The tail was grey or fawn below and on the sides, and black above, with an inconspicuous crest of longer hairs above, and a few white hairs towards the tip. The body was light and slender. The head was broad with a long, sharply-pointed snout. The ears were long and narrow. The limbs were long, slender, and peculiarly developed,

so the animal appeared to be standing on its toes, a feature that easily distinguished kantjilpa from other bandicoots, together with the low terminal crest of black hairs on the tail.

Kantjilpa occupied a wide variety of habitats in semi-arid and arid southern Australia, from the west coast at Carnarvon to western New South Wales and north-western Victoria. It made a nest of leaves lined with grass in a hollow in the ground.

Approximate date of extinction:
The last specimen was collected in about 1916 near Alice Springs. Pintupi people, however, recall it surviving in the northern Gibson Desert and southern Great Sandy Desert until the 1950s.

Probable cause of extinction:
Feral cats and foxes, combined with altered fire regimes and clearing.

Above Very few specimens of the kantjilpa exist in museums, but older western desert people recall it as being abundant until the 1950s.

Illustration – From *The Mammals of Australia* by John Gould (1863)

Walilya, desert bandicoot   *Perameles eremiana*

Other names: Orange bandicoot (English), karlkarl, kililpi, narntuurpa, nyinmi, miringinpa, nganngarrpa, tjupirrpa, walilya/warlilya, watalyari (western desert Aboriginal dialects).

Description: The walilya was similar to the marl (western barred bandicoot), but had dull orange brown fur. The head and body length was 230–285 mm and the tail length was 100–135 mm. The hair, though coarse, was not spiny. The upper parts were a grizzled grey, tinged with fawn, or almost orange brown, and the underparts were white or pale fawn. A dark stripe extended in two bands on either side of the rump. The tail was grey or fawn below and on the sides, and black above, with an inconspicuous crest and a few white hairs towards the tip. This species and the marl are closely related and may be the same species.

Walilya once occupied a variety of habitats in the central deserts of WA, the Northern Territory and South Australia, covering perhaps a quarter of Australia. It spent the daytime in a nest constructed of grass and other plant matter in a hollow in the ground, often under spinifex (*Triodia* spp.), and is reported to have fed on termites, ants and insect larvae.

Approximate date of extinction: The last specimen was collected in 1943 from Well 35 on the Canning Stock Route. It was last seen by western desert Aboriginal people near Lake Mackay about 1960.

Probable cause of extinction: Like many other extinct desert mammals, the walilya's extinction was probably due to predation by foxes and feral cats, exacerbated by changed fire regimes.

Yallara, lesser bilby   *Macrotis leucura*

Other names: Lesser rabbit-eared bandicoot (English), gurrawal, nantakarra, natukutiri/ngatukutiri, nyunpi, ngatukulirra, ngatukutita, parntakarra, tunpi, ungkanpan, wingnyil (western desert Aboriginal dialects).

Description: The yallara was smaller than the bilby (see p. 60) and was distinguished from it by having white fur along the entire upper tail. Though never collected by scientists in WA, oral history research in the 1980s showed that it was well known by older

western desert Aboriginal people, who recalled many details of its ecology. The species was once widely distributed through the deserts of WA, the Northern Territory, Queensland and South Australia, where it lived in burrows and ate small animals, including mice and termites, as well as roots, fruits and seeds.

Approximate date of extinction: This species was last collected in the Simpson Desert region in eastern Australia in 1931. Aboriginal people living near the Clutterbuck Hills in the Gibson Desert reported that the yallara survived there until the 1960s.

Probable cause of extinction: Foxes and cats, combined with altered fire regimes.

Moda, broad-faced potoroo   *Potorous platyops*

Other names: None. Moda was the Nyoongar name for this species.

Description: The head and body length was about 240 mm, whereas the tail length was about 180 mm. Moda had grizzled upper fur and was dusky white below. Its English name came from the broad skull. Although this species was collected only in the nineteenth century in the WA Wheatbelt and east of Albany, subfossil data suggest it was distributed through much of semi-arid south-western areas of WA and coastal South Australia. The only information on its habitat was provided in a note by John Gould quoting John Gilbert: 'all I could glean of its habitat was that it was killed in a thicket surrounding one of the salt lagoons of the interior'.

Approximate date of extinction: The last specimens were dated 1874 or 1875. It is not known how long it may have survived after that.

Probable cause of extinction:
Apparently extinct well before foxes arrived in WA and before widespread land clearing, its disappearance may have been due to predation by feral cats.

Below An inhabitant of Western Australia's Wheatbelt region, the moda was probably wiped out by feral cats more than 100 years ago.

Illustration – From *The Mammals of Australia* by John Gould (1863)

Boodie (mainland subspecies)   *Bettongia lesueur graii*

Other names: Boodie rat, Lesueur's rat-kangaroo, burrowing bettong, burrowing rat-kangaroo (English), boodi (Nyoongar), kunayuna, minilka/minirlka, mitika, nurrtu, pitikariti, purlana, purtaya, tjiliku, tjungku, walkaru, walkatju, wirlana, wilyinpa, yalanmunku, yiilkita, yilyikarra, yungkupalyi/yunkupayi (western desert Aboriginal dialects), yalva (Aboriginal, Broome area).

Description: Though it was similar to the Shark Bay boodie (see p. 63), the mainland subspecies was slightly larger and had a white tail tip. It was once one of the most common and widely distributed mammals in Australia, occurring in WA south of the Kimberley, southern Northern Territory, most of South Australia, western Victoria and western New South Wales. It inhabited

burrows, usually in clumped warrens with numerous entrances. Old warrens are still visible in WA deserts, especially the Gibson Desert. In May 1986, Phil Fuller and I counted boodie warrens while driving from Sandy Blight Junction (23°12'S, 129°33'E) in the Northern Territory to Gary Junction (22°30'S, 125°15'E) and then to Windy Corner (23°34'S, 125°11'E) in WA. Abandoned warrens were visible in almost all areas with hard soils (loams, lateritic surfaces, calcrete and even buckshot sandplains). Sand dunes and sandplains showed no evidence of old warrens, although they would once have occurred there. Where they were still visible, warren density was $5.9\pm0.96/km^2$. In one 5 km strip, 21 warrens were counted (equivalent to $28/km^2$). In many areas, such as the Nullarbor Plain, inland salt lakes and the south-west, former boodie warrens have been taken over by rabbits.

Approximate date of extinction: The boodie had disappeared from the south-west by the 1940s, soon after the arrival of the fox. Pintupi people reported that it survived in the Gibson Desert until about 1960.

Probable cause of extinction: Its disappearance from different parts of Australia coincided with the spread and establishment of foxes.

Nullarbor dwarf bettong   *Bettongia pusilla*

Description: Known only from subfossil remains in caves in the Nullarbor Plain, the Nullarbor dwarf bettong, as its name suggests, was smaller than other bettongs.

Approximate date of extinction: Unknown. Amy Crocker, a long-time resident of Balladonia, reported that 'rat-kangaroos' survived on the Nullarbor until the 1920s, but these could have been either (or both) the Nullarbor dwarf bettong or the woylie (*Bettongia penicillata*).

Probable cause of extinction: Predation by cats and foxes. Rabbits, which were extremely abundant on the Nullarbor Plain, would also have denuded its habitat, competed with it for burrows and/or reduced the availability of its food.

Kuluwarri, central hare-wallaby   *Lagorchestes asomatus*

Other names: Kalanpa, kulkuma, nantjwayi, pilakarratja, raputji, tjinapawulpa, tjuntatarrka, yamarri (western desert Aboriginal dialects).

Description: The kuluwarri is known to science only as a single skull in the South Australian Museum, collected in 1931 in the Northern Territory near Lake Mackay, adjacent to the WA border. Older western desert Aboriginal people remember an animal that was almost certainly this species. They describe it as a small wallaby of similar size to the boodie (*Bettongia lesueur*) with long, soft, grey fur, long hair on top of its feet and a relatively short, thick tail. It inhabited sandplains and dunes in a large area of the Great Sandy, Little Sandy, Gibson and Tanami deserts, the Central Ranges region and the eastern Pilbara.

Approximate date of extinction: This species was last seen by Aboriginal people in the southern Great Sandy Desert in about 1960.

Probable cause of extinction: Like many other extinct desert mammals, its disappearance was probably due to a combination of predation by foxes and feral cats and changed fire regimes.

Below A single skull in the South Australian Museum is the only scientific record of the kuluwarri.

Photo – Courtesy of the South Australian Museum

Woorap, rufous hare-wallaby (south-west WA)
Lagorchestes hirsutus hirsutus

Other names: Western hare-wallaby, whistler, mala (the last name, used by western desert Aboriginal people, is applied primarily to the central Australian subspecies). Woorap is the Nyoongar name, variously recorded as woorup, wurrup, wurup, woorark and wirrup by early European authors.

Description: This small wallaby reached about 350 mm in body length with a 250–300 mm tail. It had long, reddish-brown fur above, and paler fur below. It inhabited sandplains with low woody shrubs, in the Wheatbelt and adjacent areas, and was herbivorous. Hare-wallabies are so-named because of their habit of sitting very still under a shrub or in a shallow scrape and, when approached closely, exploding from the hide in a similar manner to European hares.

Approximate date of extinction: The last specimen from the south-west came from York and was collected in 1846. Bruce Leake, who lived near Kellerberrin, reported that it disappeared from that area between 1894 and 1899. It probably became extinct soon afterwards.

Probable cause of extinction: Predation by feral cats.

Maning, banded hare-wallaby (mainland) *Lagostrophus fasciatus albipilis*

Other names: Maning is the Nyoongar name, recorded by some early authors as mernine, munning and marnine.

Description: The maning had a head and body length of 400–450 mm, with a 350–400 mm long tail. It was a dark grizzled grey above, with transverse dark bands across the lower half of its back and rump. The subspecies once occurred at Shark Bay (possibly including Dirk Hartog Island) through the semi-arid areas of the south-west, south-east to near Esperance and eastwards to the Nullarbor Plain. Recently, a specimen from near Adelaide was located in a museum in Berlin. Subfossil records from the Narracoorte Coastal Plain of western Victoria show that it occurred more widely in south-eastern Australia.

The taxonomy of subspecies within *L. fasciatus* is unclear. A recent paper suggests that all WA animals (including those from Bernier and Dorre islands) belong to *L. fasciatus fasciatus* and eastern animals to a new subspecies *L. fasciatus baudinettei*.

Approximate date of extinction: The last records were in the south-west of WA, where it probably became extinct early in the twentieth century. Bruce Leake, who lived near Kellerberrin, reported that it disappeared from that area between 1894 and 1899. Guy Shortridge collected it east of Pingelly in 1906. It is not known how long it survived afterwards.

Probable cause of extinction: Predation by feral cats, land clearing and competition for food by rabbits probably all played a role.

Tjawalpa, crescent nailtail wallaby *Onychogalea lunata*

Other names: Worong (Nyoongar), kurrungku/kurungku/kurrurungku, pitarri, tjawalpa wamarru, warlpatju/walpatju (western desert Aboriginal dialects).

Description: This medium-sized wallaby had a body length of about 400–500 mm and a tail up to 330 mm long. It was generally ash grey above, pale grey below, and had defined crescent-shaped white shoulder stripes, ill-defined hip stripes and a horny 'nail' at the end of its tail. It extended from the WA Wheatbelt through much of the southern deserts and lived mainly in woodlands, including those of eucalypts and mulga (*Acacia aneura*).

Approximate date of extinction: It was last reported from the western deserts in the 1940s or 1950s.

Probable cause of extinction: Predation by foxes, possibly exacerbated by changed fire regimes, seems the most likely cause of extinction.

Below An inhabitant of woodlands, the beautiful tjawalpa did not survive the arrival of foxes.

Illustration – From *The Mammals of Australia* by John Gould (1863)

Djooyalpi, lesser stick-nest rat *Leporillus apicalis*

Above Although stick-nest rats disappeared from mainland Australia in the 1930s, their stick-and-stone nests can still be found in some areas, such as this nest in the Young Range in the Gibson Desert.

Photo – David Pearson/CALM

Other names: White-tipped stick-nest rat (English), tjuyalpi, yinunma/yinurnma/yininma (western desert Aboriginal dialects). The Aboriginal names may apply to either or both species of stick-nest rats.

Description: The djooyalpi had a head and body length of about 230 mm, and a tail about 230 mm long. It weighed about 125 g. It was smaller and more slightly built than the wopilkara (*L. conditor*) and was distinguished by long white hairs on the last quarter of its tail. It was once very widespread through most of semi-arid and arid country in WA, the Northern Territory and South Australia, and occurred in most of WA apart from the Kimberley and the extreme south-west corner. Djooyalpi extended further northwards than wopilkara (see p. 73).

Approximate date of extinction: Both species of stick-nest rats became extinct on the mainland in the 1930s. The large stick-and-stone nests of this species and/or wopilkara can still be seen in caves, particularly in the Gibson Desert. They may be more than 2 m in diameter.

Probable cause of extinction: Predation by feral cats and foxes.

Yoontoo, short-tailed hopping-mouse *Notomys amplus*

Description: The yoontoo differed from other hopping-mice by having a short tail and large ears, which were nearly as long as its head.

Approximate date of extinction: Only two specimens of the yoontoo were ever collected, both at Charlotte Waters in northern South Australia in 1896.

Subfossil data show that it had a widespread distribution in arid Australia, from the west coast near Carnarvon through the deserts eastwards to the Flinders Ranges and the Simpson Desert in South Australia.

Probable cause of extinction: Predation by feral cats.

Noompa, big-eared hopping-mouse *Notomys macrotis*

Other names: Large-eared hopping-mouse (English), bolong (Nyoongar).

Description: This large hopping-mouse of about 60 g was greyish-brown above and white below. It had large ears.

Approximate date of extinction: The noompa was collected only at New Norcia prior to 1844. Subfossil data have not been able to extend its former range beyond the northern Wheatbelt.

Probable cause of extinction: Predation by feral cats, possibly exacerbated by damage to burrows and competition from sheep and other hard-hooved introduced animals.

Koolawa, long-tailed hopping-mouse *Notomys longicaudatus*

Other names: Koolawa, kodong (Nyoongar).

Description: This large hopping-mouse had a head and body length of about 150 mm and a crested tail 150–200 mm long. It probably weighed about 100 g. Its upper fur was reddish-brown, and it was white below.

Approximate date of extinction: Like the noompa, the koolawa was first collected at New Norcia prior to 1844, however, it was also collected near Broken Hill in New South Wales, and near Alice Springs in the Northern

Territory. Subfossil data show it had a wide distribution through WA south of the Kimberley, southern Northern Territory, western New South Wales and western Queensland. In WA, it occurred in the Wheatbelt, Geraldton sandplains, Murchison, Carnarvon Basin, and in the Great Sandy, Little Sandy, Gibson and Great Victoria deserts. The last specimen was collected in 1901 and the koolawa probably became extinct soon afterwards.

Probable cause of extinction: Feral cats. Changed fire regimes, rabbits and stock may have contributed to its demise.

Extinct in the wild subspecies

Top A mala being released on Trimouille Island in 1998.

Photo – Andrew Burbidge /CALM

Above Central Australian mala have the rufous colour that gives the animal its English name—rufous hare-wallaby.

Photo – Stanley Breeden

Mala *Lagorchestes hirsutus* undescribed central Australian subspecies

Other names: Rufous hare-wallaby (mainland), brown hare-wallaby, spinifex rat (English name also applied to other desert wallabies and bandicoots), atnukwa, irlraku, kunatjinpa, landaa, landalyparti, liwilpa, mala, malyi, matjiri/matjirri, ngartinpa, ninngka, parranti, pilki-pilki, raltatu, tarnnga, tintinpa, tipirri, tiwilpa/liwilpa, tjanpingkatja, tjanpitja, tjiwilpa/tiwilpa, tjunpu, warku, wirrini, witjari/witjarri (western desert Aboriginal dialects).

Description: Mala have a head and body length of 310–360 for males and 360–390 for females, with a tail of 260–300 mm. Weight is 1250–1800 (males) or 780–1960 (females). They are rufous above and paler below, with a dark grey or rufous head. They hold their forelegs widely separated when hopping.

Distribution and habitat:
This subspecies was once very widespread and common in the Great Sandy, Little Sandy, Gibson, Tanami and Great Victoria deserts, the Central Ranges region, central and southern Northern Territory and north-western South Australia. By the 1950s it was considered extinct. Then, in 1959, Alan Newsome located a small colony on the Tanami Stock Route, about 450 km north-west of Alice Springs. Subsequent surveys discovered another small colony about 10 km away near Sangsters Bore. These two colonies were lost in 1987 and 1991, one by fox predation and the other by fire. Fortunately, a breeding colony had been established in Alice Springs, and further captive colonies have been established from it.

Biology and ecology: Mala inhabit spinifex hummock grasslands, sandplains and dunes, and feed mainly on spinifex seed heads. The young leaves of sedges and grasses are also eaten, along with herbs and shrubs. Breeding has been recorded throughout the year, but the timing depends on rainfall. Gestation is probably less than four weeks and pouch life is 14-16 weeks. Females start breeding at about nine months of age. Under ideal conditions, three young can be raised in a year.

Threats: A colony introduced to Trimouille Island, in the Montebello Islands, could be threatened by introductions of predators or extensive fire. Mainland reintroductions are very unlikely to be successful unless feral cats are nearly eliminated.

Status: Extinct in the wild. Attempts to reintroduce mala to the wild in the Tanami Desert, Northern Territory, failed due to feral cat predation. In June 1998, 30 mala, 14 with pouch young, were translocated to Trimouille Island from a semi-captive colony in the Tanami Desert. Most of these animals survived and bred, and the population is expanding. Reassessment of the mala's conservation status should soon be possible.

Captive-bred mala, also from the Tanami Desert, are being captive-bred at Peron Peninsula in Shark Bay ('Project Eden') and Dryandra Forest in the Wheatbelt ('Return to Dryandra'). Some animals were released into the wild on Peron Peninsula in 2001, but several were killed by feral cats and the rest were returned to captivity pending better cat control.

Threatened species and subspecies

Chuditch   *Dasyurus geoffroii*

Other names: Western native cat, western quoll (quoll is an Aboriginal name from Queensland for the northern quoll *D. hallucatus*), djooditj, badjada, ngooldjangit (Nyoongar), gnangnalpa, kingkin, kinkilpa, kitjikurna, kunatjirila, kurninngka, maularurru, ngal-ngal/ngarl-ngarl, parrtjarta, tjalpartu, tjatjirti, walyparti, wiminytji/wimitji, yunanngnalku (western desert Aboriginal dialects). Wunambal Aboriginal people from the northern Kimberley report a large *Dasyurus* from there, which they call dada, so chuditch (or another similar species) may occur in that area.

Description: The chuditch is the largest carnivorous native mammal in WA and the only spotted mammal south of the Pilbara, where the northern quoll occurs. Adults are about 250–400 mm long (head and body), with a 200–350 mm long tail, and weigh 1–2 kg. Males are larger than females. Chuditch have rufous grey upper fur with white spots and pale creamish-grey underparts.

Distribution and habitat: Chuditch once occurred throughout the State, excluding the Kimberley, occupying a wide variety of habitats. By the 1970s, they were restricted to the jarrah forests of the south-west and a few larger Wheatbelt reserves, with scattered outlying populations in the eastern Goldfields.

Biology and ecology: Chuditch consume a wide range of prey, mostly eating small vertebrates and larger invertebrates. They breed in winter, with juveniles dispersing in early summer.

Threats: Predation by foxes is a proven threat. An experiment carried out at Batalling Forest, north of Collie, by Keith Morris and colleagues from the Department of Conservation and Land Management showed conclusively that fox control resulted in increased chuditch recruitment and survival. Chuditch numbers in the jarrah forest and some other sites have increased significantly since *Western Shield* (see Chapter 3) was implemented. Chuditch have been translocated to Kalbarri National Park, Julimar Conservation Park (between Toodyay and Bindoon), the proposed Mt Lindesay national park (near Denmark), Lake Magenta Nature Reserve (in the Wheatbelt) and Cape Arid National Park (east of Esperance).

Status: Vulnerable. A ten-year recovery plan published in 1991 has been successfully implemented. The chuditch may qualify for removal from the list of threatened fauna before long.

Below Once seemingly headed for extinction, the chuditch is now recovering following fox control and translocations.

Photo – Babs and Bert Wells /CALM

Above and below Very common in sandy deserts until the 1930s or 1940s, mulgaras now survive in small scattered colonies.

Photos – Babs and Bert Wells /CALM

Mulgara *Dasycercus cristicauda*

Other names: Kakati, kurrakura, miniilka, minirlya, minyiminyi, nyarlurti, murrtja, munyantji/minyantji, papanytji/papalytji, putjapurru, talingkatawun, tjatjalpi (western desert Aboriginal dialects). These names may also apply to the ampurta (see opposite), or some may be names for the ampurta.

Description: This solid, muscular marsupial has a body length of about 125–220 mm and weighs from 60 g to 130 g (males are larger). It has grey to rufous fur above and greyish-white fur below. The tail is about 75–120 mm long. It is distinguished from the ampurta by the crest on its tail, which starts one third of the way from the tip on the upper surface. The crest consists of shorter hairs at the tip progressing to longer ones. The females have six nipples.

Distribution and habitat: Formerly widespread in sandy deserts, mulgaras are now rare and patchily distributed. In WA, they have recently been recorded from the Great Victoria, Gibson, Great Sandy, Little Sandy and Tanami deserts, the Pilbara, Gascoyne, Murchison, north-eastern Goldfields, the Central Ranges region and Carnarvon Basin (Kennedy Range). The mulgara constructs burrows in spinifex (*Triodia*) hummock grassland.

Biology and ecology: Mulgaras eat small vertebrates and larger invertebrates. They breed in winter, producing four to eight young, which become independent by spring. The species lives primarily in relatively long-unburnt habitat.

Threats: Predation by foxes and cats, inappropriate fire regimes.

Status: Vulnerable. Recovery work in the Northern Territory, South Australia and WA is coordinated by a recovery team. Research into its taxonomic status has confirmed that the ampurta is a separate species.

Ampurta   *Dasycercus hillieri*

Other names: Western desert Aboriginal names for the mulgara (see p. 48) may also apply to this species.

Description: The ampurta is similar to the mulgara, but is slightly larger and paler in colour. The crest on its tail consists of black hairs, which are all about the same length, starting in the middle of the upper surface of the tail. Females have eight (occasionally seven) nipples.

Distribution and habitat:
The ampurta was originally distributed throughout the Great Sandy, Little Sandy, Gibson, Tanami and Simpson deserts, the Channel Country in south-west Queensland and the Nullarbor Plain, and possibly in areas in between. This species was found along the Canning Stock Route in about 1930, but no specimens have been collected in WA since 1947, when it was found at the Blue Spec Mine near Nullagine. A survey in 2002—undertaken by the Arid Land Environment Centre, the Department of Conservation and Land Management and local Aboriginal people around Wells 28 to 33 on the Canning Stock Route—resulted in the capture of several mulgaras but no ampurta. Recent surveys in the Simpson Desert in South Australia have revealed that it is relatively common there.

Biology and ecology: Studies into the ecology and conservation of the ampurta are being carried out in the Simpson Desert, where the species appears to be more common than it is in WA.

Threats: Ampurta are probably threatened by predation by feral cats and foxes, and possibly by habitat degradation resulting from changed fire regimes.

Status: Endangered. The conservation of the ampurta is being addressed by the same recovery team as the mulgara. Further searches for ampurta in WA should be undertaken.

Below The ampurta has not been collected in Western Australia since 1947.

Photo – Peter Canty /SA Department for Environment and Heritage

Above Dibblers have distinctive white eye rings.

Below Long thought to be extinct until its rediscovery in 1967, the dibbler now appears to have a more secure future. The recovery plan for this species involves research, captive breeding and translocations.

Photos – Babs and Bert Wells /CALM

Dibbler   *Parantechinus apicalis*

Other names: Dibla, madoon (Nyoongar).

Description: Dibblers have a head and body length of 140–145 mm. The 95–115 mm long tail is thickened at the base, tapering and covered with dense hairs. Females weigh 40–75 g and males 60–100 g. They are brownish-grey above, freckled with white, and greyish-white below, with a yellowish tinge where the dark upper fur meets the light underbody fur. There is a distinct white eye ring.

Distribution and habitat: Originally collected at New Norcia, the dibbler was once apparently widespread in the northern kwongan (shrubby heathlands) from Shark Bay (including Dirk Hartog Island) south to near Perth, in the southern kwongan from Albany to Israelite Bay and on the Eyre Peninsula, South Australia. The species was long thought to be extinct until found in 1967 by Michael Morcombe at Cheyne Beach near Albany. Since its rediscovery, populations have been located in Torndirrup National Park near Albany and Fitzgerald River National Park to the east. Specimens have also been found east of Fitzgerald River National Park near Hopetoun and Jerdacuttup. In 1985, a Department of Conservation and Land Management survey located populations on Boullanger (34 ha) and Whitlock (7 ha) islands in Jurien Bay. Dense populations of house mice on both islands may be affecting dibbler numbers through competition for food. No technique currently exists to eradicate the mouse populations without affecting the dibbler (and Boullanger Island dunnart) populations.

Biology and ecology: Dibblers are insectivorous and, unlike most WA native mammals, very active during the day. They breed in March, producing up to eight young. Juveniles become independent at three to four months and breed at 11 months. Dibblers may live for about four years.

Threats: Land clearing and predation by cats were major threats in the past. Current threats are cats and inappropriate fire regimes. Island populations are threatened by the potential introduction of predators, including rats.

Status: Endangered. The dibbler recovery team is implementing a recovery plan involving study of mainland and island populations, captive breeding at Perth Zoo and translocations. Captive-bred dibblers, derived from animals from Boullanger and Whitlock islands, have been successfully introduced to the mouse-free Escape Island in Jurien Bay. In 2001, 41 captive-bred dibblers, of mainland origin, were released in the proposed Peniup nature reserve between Fitzgerald River and Stirling Range national parks. Four released dibblers were killed by grey currawongs (*Strepera versicolor*), showing the importance of dense, long-unburnt vegetation for this species on the mainland. Monitoring suggests that this translocation will be successful.

Kingo, red-tailed phascogale
Phascogale calura

Other names: Ngintingkaparrtjilaralpa, papalakurntalpa, papatjakurnalpa (Pintupi). Kingo is the Nyoongar name for this species.

Description: Females have a head and body length of 93–105 mm, tail length of 119–144 mm and weigh 38–48 g, whereas males are 105–122 mm long, with 134–145 mm long tails and weigh 39–68 g. Kingo are ash grey above (darker in front of the eye) and cream to white below. The ears and the base of the tail are reddish, and the outer half of the tail has a brush of long black hairs. Kingo are distinguished from wambenger (an undescribed species of brush-tailed phascogale from the south-west of WA) by their reddish tail base.

Distribution and habitat: Kingo were once widely distributed from the south-west of WA to the Great Sandy Desert. They also occurred in south central Northern Territory, southern South Australia, the Flinders Ranges and on the Victoria-New South Wales border near Darling Junction. The species had probably disappeared from the eastern states by the late 1800s. The last western desert specimen was collected from Well 44 on the Canning Stock Route in 1931, although western desert Aboriginal people believe that kingo may persist in the southern Great Sandy Desert. Small, scattered populations still occur in remnant vegetation in the Wheatbelt.

Biology and ecology: This insectivorous and carnivorous marsupial consumes birds, small mammals, insects and spiders, and spends most of its time in trees. It shelters mainly in eucalypt hollows, including those of sand-dune bloodwood (*Eucalyptus chippendalei*) on dunes in the Great Sandy Desert, but will also use grasstree (*Xanthorrhoea*) leaf clumps and sheoaks (*Allocasuarina*). In the south-west, mating occurs in winter and most (probably all in the wild) males die soon afterwards. Juveniles become independent by spring and breed when less than a year old.

Threats: Feral cats, inappropriate fire regimes. Most reserves in which the kingo remains have not been burnt for a long time. Fire research at Tutanning Nature Reserve, in the Wheatbelt, showed that some kingo survived a low-intensity fire.

Status: Endangered. Kingo occur in several, mostly small, Wheatbelt reserves including Dryandra Forest and Tutanning, Boyagin, North Karlgarin, Bendering, Dongolocking, Pingeculling, East Yornaning and Yilliminning nature reserves, and other small remnants. Their status is being monitored under *Western Shield*. Fire management is an important component of their conservation, as fire can open up dense habitats and thus allow increased predation by cats.

Above Formerly widespread in the western deserts, the kingo is now restricted to remnant vegetation in WA's Wheatbelt.

Photo – Babs and Bert Wells /CALM

Above A Butler's dunnart.

Photo – Damian Milne /Department of Infrastructure, Planning and Environment, Darwin

Below The Boullanger Island dunnart inhabits low scrubby vegetation.

Photo – Andrew Burbidge /CALM

Butler's dunnart *Sminthopsis butleri*

Other names: Ungaringyin Aboriginal people call this and other small mammals munjol and the Wunambal word for small mammals is dadaru.

Description: Butler's dunnart has a head and body length of 88 mm and a thin and sparsely-haired tail of around 90 mm long. Its soft fur is greyish above, with an indistinct head stripe, and white below. Butler's dunnart is readily distinguished from the red-cheeked dunnart (*S. virginiae*), which has rufous fur on the sides of its head.

Distribution and habitat: In WA, this species was known only from near Kalumburu, where it was first collected by Harry Butler in December 1965 on a black-soil plain and in flood debris near King Edward River. It has not been found since in WA. Butler's dunnart also occurs on Bathurst and Melville islands in the Northern Territory.

Biology and ecology: Very little is known about Butler's dunnart. One of the original specimens, a female, had seven small pouch young.

Threats: Inappropriate fire regimes, damage to habitat by stock, wild cattle and donkeys, and predation by feral cats are likely threats.

Status: Vulnerable. Further surveys to locate populations are necessary before conservation actions can be designed.

Boullanger Island dunnart *Sminthopsis griseoventer boullangerensis*

Description: Boullanger Island dunnarts are light greyish-olive above and light grey below. The feet are white. The tail is longer than the head and body length (68–88 mm). In the mainland subspecies the tail is slightly shorter. Females weigh 9–15 g and males weigh 14–17 g.

Distribution and habitat: This dunnart is known only from Boullanger Island in Jurien Bay. However, a specimen from Lesueur National Park, and subfossil material from caves inland from Jurien, may also belong to the subspecies, so further survey and taxonomic research are needed to clarify its status and range. The grey-bellied dunnart (*S. griseoventer griseoventer*) occurs from near Geraldton to Lake Magenta and near the south coast east to Esperance, with an outlying population on the Roe Plain.

Biology and ecology: Research suggests there are between 100 to more than 300 animals on Boullanger Island, depending on time of year and seasonal conditions. Mating is in July. Females carry litters of up to eight young in August. Pouch life lasts 4–5 weeks. The young are then deposited in a leaf-lined nest just under the soil surface and emerge at 10 weeks of age. Young females remain near their area of birth, but males disperse up to several hundred metres within two months of weaning. Survival of both sexes up to this time is less than 50%. Females produce only one litter a year and occasionally produce young in two consecutive seasons. Males and females become sexually mature at about one year and live for a maximum of 2¹/₂ years.

Threats: A dense population of house mice on Boullanger Island may affect the dunnarts by competing for food. Dibblers (see p. 50) also occur on Boullanger. No techniques currently exist to eradicate the mouse population without affecting the dibbler and dunnart populations.

Status: Vulnerable. The dunnart population is being monitored.

Sandhill dunnart
Sminthopsis psammophila

Description: These dunnarts have a head and body length of 85–114 mm, a tail 107–128 mm long and weigh 30–55 g. Males are larger than females. They have a pale grey head, black eye rings, and buff cheeks and flanks. The upper fur is grey to buff, with white underparts and feet. The tail is pale grey above and dark grey below, tapers towards the tip and has a distinctive crest of blackish-grey hairs on the terminal quarter. It is the second largest dunnart. Only the Julia Creek dunnart (*S. douglasi*) from Queensland is as large or larger.

Distribution and habitat: This species was first collected near Lake Amadeus in the Northern Territory (NT) in 1894, but has not been recorded in the NT since. In recent decades it has been found in the Great Victoria Desert in WA and South Australia (SA), and the Eyre Peninsula in SA. Subfossil specimens show that it once occurred in the Great Sandy Desert, in the Murchison, in the southern NT, and in the Gawler Ranges area of SA. It is probably locally extinct in the Eyre Peninsula. In WA, it has been found in and to the north of Queen Victoria Spring Nature Reserve.

Biology and ecology: Sandhill dunnarts inhabit sandy soils with spinifex, in areas of low open woodland and scattered mallees and shrubs. They can traverse long distances in a single night. One male was recorded covering 1960 m in just two hours. They appear to have reasonably stable home ranges and shelter by day in spinifex, logs, hopping-mouse burrows or burrows dug between spinifex hummocks. The diet is largely insects and spiders, with small amounts of vertebrate prey. Mating occurs in September, with young born in September or October. Pouch young are weaned in December or January, but newly-independent juveniles have been

Above An inhabitant of the Great Victoria Desert, the sandhill dunnart is rarely recorded and poorly known.

Photo – Babs and Bert Wells /CALM

found in October and April, suggesting that in good seasons they can alter the timing or duration of breeding.

Threats: Threats include predation by cats and foxes and changed fire regimes. Land clearing has destroyed the sandhill dunnart's habitat on the Eyre Peninsula.

Status: Endangered. A draft national recovery plan was prepared by the SA Department of Environment and Conservation in 2001, but implementation has not commenced. Studies by WA Department of Conservation and Land Management staff near Queen Victoria Spring continued until March 2003, when a wildfire burnt the area out. Captures of sandhill dunnarts were sporadic (17 over 13 years) and only one male was caught on the last trapping trip in March 2000. Recent surveys located a population approximately 50 km north of Queen Victoria Spring. Further surveys are needed to locate populations that can be studied so conservation requirements can be determined. Fire management is important, as sandhill dunnarts inhabit areas of mature spinifex, but abandon areas that remain long unburnt.

Numbat *Myrmecobius fasciatus*

Other names: Noombat, wioo (Nyoongar), mutjurarranypa, parrtjilaranypa, walpurti (western desert Aboriginal dialects). Numbat is mispronounced by most Australians, who pronounce the 'u' as in 'hunt' and 'a' as in 'bat', rather than 'oo' as in book and 'a' as in 'media'.

Description: The numbat cannot easily be confused with any other mammal. It has a head and body length of 200–274 mm, a tail length of 160–210 mm and weighs 300–700 g (mean 470 g). It is reddish-brown above, changing to black at the rear, with a series of prominent transverse white stripes. The head is narrow with a pointed snout and a horizontal eye stripe, and the tail is long and hairy. Numbats have 50–52 teeth, but these are poorly developed and not used in eating.

Distribution and habitat: The species was once widespread across southern Australia from the west coast to central New South Wales and north-western Victoria. It occurred in the southern Great Sandy Desert, the Gibson Desert and the Great Victoria Desert. Pintupi people in the Gibson Desert remember numbats occurring there until about 1965. By the early 1970s, the numbat was restricted to drier parts of WA jarrah forests and some Wheatbelt remnants, and by the late 1970s there were serious concerns that it was on the verge of extinction.

Biology and ecology: The numbat is active only during the day. It feeds solely on termites, for which it digs with its long, sharp claws and gathers with its long, sticky tongue. In the south-west it usually shelters in hollow logs on the ground, in tree stumps or in low hollow tree limbs. In drier areas without trees it sheltered in a burrow. Females have four young, born in mid to late summer, that become independent by October. Once the young are too large to be carried on the teats, the mother deposits them in a chamber, lined with grass and leaves, at the end of a narrow burrow.

Threats: Predation by foxes and cats. Birds of prey, especially goshawks but also larger species such as wedgetail eagles, are natural predators. Clearing and habitat fragmentation have probably increased the numbers of some of these predators in remnant vegetation, putting additional pressure on the remaining small numbat populations.

Status: Vulnerable. The conservation of the numbat—Western Australia's mammal emblem—has been a high priority of the Department of Conservation and Land Management and its predecessors since the late 1970s, when it became clear that numbers had crashed. A recovery plan, first developed in the early 1980s, has been implemented and updated since. Fox control, first at Dryandra State Forest (a proposed national park) and later elsewhere, led to significantly increased numbers, but numbers within most populations have since declined (though not to the same levels as before fox control), perhaps because of cat predation. Perth Zoo has bred numbats in captivity and both wild-caught and captive-bred numbats have been translocated to parts of the jarrah forest (Batalling and Dale), and to Boyagin, Karroun Hill and Dragon Rocks nature reserves and Stirling Range National Park. Captive-bred animals survive better when released if trained to avoid birds of prey. Even after all this conservation work, the numbat is still a very rare species.

Above Juvenile numbats at Dryandra Woodland. Females have four young in each litter.

Opposite Numbats were once widespread across much of southern Australia. They are now naturally restricted to the south-west, where they often shelter in hollow logs.

Photos – Babs and Bert Wells /CALM

Kakarratul, northern marsupial-mole   *Notoryctes caurinus*

Above A marsupial mole.

Photo – D Roff/NatureFocus

Below A kakarratul from northern Western Australia.

Photo – Andrew Burbidge /CALM

Other names: Kakarratul/kakarratulpa, ngarrinatjulurru/ngarrinytjatjulurru, pikarnawarlartarri, tjilpirrkuwakanytja (western desert Aboriginal dialects), 'pensioner' (western desert Aboriginal slang, so-called because it has whitish hair and is blind). Some of the Aboriginal names may apply to either or both species of marsupial-mole.

Description: Marsupial-moles are small (with a head and body length of about 130–150 mm), compact animals weighing 40–70 g. No eyes are apparent (they are reduced to non-functioning lenses beneath the skin) and the ears are reduced to holes covered by dense fur. They eat insects, especially larvae and pupae. The kakarratul has orange fur, while the itjaritjari is off-white.

Distribution and habitat: The kakarratul occurs in the Great Sandy Desert and adjacent pindan country to the north, the Little Sandy and northern Gibson deserts, and the central ranges region.

Biology and ecology: The kakarratul spends most of its life underground, but occasionally comes to the surface or moves just below the surface, perhaps more frequently after rain, at which times it is vulnerable to predation.

Threats: Predation by foxes, cats and dingoes is the main threat.

Status: Endangered. The range of the species does not seem to have contracted, but there is abundant testimony from Aboriginal people and other sources that the kakarratul has become much less common. Research into its conservation biology is urgent.

Itjaritjari, southern marsupial-mole   *Notoryctes typhlops*

Other names: Itjarritjarri/itjarri-itjarri, kakarratulpa/kakarrartuunpa, yitjarritjarri, yirtarrutju (western desert Aboriginal dialects), 'pensioner' (see kakarratul above).

Description: See kakarratul.

Below Itjaritjari.

Photo – Mike Gillam

Distribution and habitat: The itjaritjari is found in the southern Gibson and Great Victoria deserts and the central ranges region. It also occurs in central northern South Australia and the southern Northern Territory.

Biology and ecology: See kakarratul.

Threats: Predation by foxes, cats and dingoes.

Status: Endangered. Like that of the kakarratul, the range does not seem to have contracted, however, it has become much less common, probably more so than its close relative. A study being undertaken with Aboriginal people in north-western South Australia will hopefully provide information on its biology and ecology, and indicate what needs to be done to conserve it.

Wintarru, golden bandicoot
Isoodon auratus auratus

Other names: Kantjalpa/kantjarrpa, lutalpa, makurra, manparri/manpayarri, minawarrulpa, minganypa, mingatjurru/minantjurru, mingawarrulpa, mirtara, mulyatjuku/mulyatjurru, mulyu, pakuru, parranpalyi, piilkarra, tirinpa, tirrka/tirrika, tjilpuku, tjirika, tjulpuku, tjurrungu, ulkaratja, wintarru/wirntarru (western desert Aboriginal dialects), gulngan? (Bardi, Kimberley), kranpula (Ungaringyin, Kimberley), garimbu (Wunambal, Kimberley).

Description: Wintarru have a head and body length of 200–250 mm, a tail 100–110 mm long, and weigh 300–650 g. They are similar to quenda of the south-west (recent genetic research suggests they may be the same species, with wintarru being a distinct subspecies), but their hair is tinged golden brown on the back and sides. The brindled bandicoot (*Isoodon macrourus*), also found in the Kimberley, has a head and body length of up to 400 mm, weighs up to 2100 g, and lacks the golden brown colour. However, separating young brindled bandicoots from young wintarru can be very difficult.

Distribution and habitat: Wintarru once occurred over about a third of Australia, including the Kimberley, Great Sandy, Little Sandy and Gibson deserts, most of the Northern Territory and northern South Australia. In the western deserts it was very common and a major food item for Aboriginal people, as reflected in the large number of names for it in western desert dialects. The last desert specimen was taken in 1952 at the Granites in the Tanami Desert. Aboriginal people living near Lake Mackay in the Great Sandy Desert stated that it survived there until about 1965–70.

Small remnant populations still occur in near-coastal areas of the north-west Kimberley, and on Augustus and Uwins islands. Wintarru also occur on Marchinbar Island in the Northern Territory. The populations on Barrow and Middle islands are treated as a separate subspecies (see p. 58).

Biology and ecology: Wintarru lived in a wide variety of habitats, including sandplains and dunes, savannah woodland, spinifex (*Triodia*) on rugged sandstone and along the edges of vine thickets. During the day, they shelter in spinifex hummocks, dense grass and rock piles. They may construct a nest.

Threats: The disappearance of wintarru from the deserts coincided with the arrival of the fox and major changes in fire regimes as Aboriginal people moved to settlements. Remnant populations in the Kimberley are threatened by too frequent and extensive fires and by predation by feral cats.

Status: Vulnerable. Better fire management is urgently needed in the Kimberley. Implementation of broadscale feral cat control would assist this and many other species.

Above Wintarru, here photographed in the Kimberley, are closely related to quenda (or southern brown bandicoot) in the south-west of WA.

Photo – Tricia Handasyde /CALM

Barrow Island golden bandicoot *Isoodon auratus barrowensis*

Description: Smaller than mainland wintarru, Barrow Island golden bandicoots have a head and body length of 190–220 mm and weigh 265–500 g. Recent genetic research suggests that Barrow Island animals may not be in a separate subspecies to mainland golden bandicoots, and that the golden bandicoot may be a subspecies of the quenda (*I. obesulus*).

Distribution and habitat: Golden bandicoots are very common on Barrow Island, occupying all habitats, and also occur on nearby Middle Island. They are extinct on Hermite Island, the largest of the Montebello Islands.

Biology and ecology: Females give birth to up to six young, but only one or two survive. Young may be born throughout the year, with the timing dependent on rainfall. Like other bandicoots, Barrow Island animals are omnivorous, eating small animals such as termites and a variety of plant material. Turtle eggs are a favourite food during the turtle nesting season.

Threats: Introduction of feral predators or disease.

Status: Vulnerable. 'Montebello Renewal', a *Western Shield* project, is eradicating feral cats and rats from the Montebello Islands. Once this is achieved, Barrow Island golden bandicoots can be reintroduced to Hermite Island, to establish a third population of this subspecies. A 1992 translocation to an area where dingoes and foxes had been eliminated in the Gibson Desert Nature Reserve failed due to feral cat predation. Barrow Island mammals are being monitored by a joint Department of Conservation and Land Management and ChevronTexaco study.

Below Golden bandicoots are abundant on Barrow and Middle islands, off the Pilbara coast.

Photo – Babs and Bert Wells /CALM

Western barred bandicoot (Shark Bay islands)
Perameles bougainville bougainville

Description: Western barred bandicoots have a head and body length of 200–300 mm and a tail length of 75–120 mm, weighing 190–250 g. They are light grey to brownish-grey above, and whiter below. Two or three alternating muted paler and darker bars are evident across the hindquarters. The feet are white and the ears are large and erect.

Distribution and habitat: Restricted to Bernier and Dorre islands, Shark Bay.

Biology and ecology: Western barred bandicoots inhabit most parts of Bernier and Dorre islands, but are especially common on white sand dunes behind beaches. During the day they occupy a nest, usually well concealed and difficult to locate, made from grass and other vegetation and placed in a hollow or under a shrub. They emerge at late dusk to feed on insects and other small animals, seeds, roots and herbs, obtained by digging or hunting. Breeding has been reported in autumn and winter, and two young are usually carried in the pouch.

Threats: Any introduction of feral predators or disease, or frequent or extensive fire, could have a devastating effect on the only two populations.

Status: Endangered. Western barred bandicoots have been reintroduced to Heirisson Prong, a 1000 ha fenced-off peninsula in Shark Bay. Western barred bandicoots sourced from Bernier Island have been captive-bred as part of 'Project Eden', which aims to reconstruct the fauna of Shark Bay within the Francois Peron National Park. However, because of the discovery of a wart-like disease in the captive animals, and because feral cat control at Peron has been less successful than was hoped, reintroduction plans have been placed on hold. Western barred bandicoots sourced from both Bernier and Dorre islands are being bred in captivity at Kanyana Wildlife Rehabilitation Centre. Eighteen animals from Kanyana have been transferred to the 'Return to Dryandra' breeding enclosure at Dryandra Woodland, where they joined animals that came direct from the islands or from 'Project Eden'. Some bandicoots were released into a small enclosure at Dryandra in 1998 and 1999, as a first step towards wild release, but the discovery of the wart-like disease in captive animals and those due for release led to the delay of reintroductions until research into the disease is completed.

Above Western barred bandicoots were restricted to Bernier and Dorre islands in Shark Bay. They are now being reintroduced to some mainland sites.

Photo – Babs and Bert Wells /CALM

Above and below The bilby has become one of Australia's best known mammals and is replacing the 'bunny', an introduced pest, as a symbol of Easter.

Photo (above) - Babs and Bert Wells/CALM

Photo (below) - Ray Smith /CALM

Bilby, dalgyte *Macrotis lagotis*

Other names: Rabbit-eared bandicoot, greater bilby, pinkie (English), dalgyte (English derived from Nyoongar), gumajin (Bardi), djalkat/djalkait (Nyoongar), atnungwa, atnunka, kartukarli, kulkanula, kulkawalu/kulkawalwal/kulkuwala/ kurkuwalwa, kunatjalku, kunatjupi, kurkawalwal/kurkuwalu, mankarrpa, maratji, marrura, minyirrika, mitulurtju, naluwarra, nanakata, nantalpa, ngakamiti, ngunpi, nini, ninu/nirnu, nirlyari/nirlari/ nilaru, nunpi, nyalku/nalku, nyunpi, pangkarru, parnpurtu, partiri, pintitiri, pirntikari, puntikulu, talku/tjalku, tjapatu, tjika, tjilikun, tjirratu, walpatjirri, wartikiti, witji, yalpaurini, yalwirti/yalwari, yariningi, yinpu/yinpuru (western desert dialects).

Description: Female bilbies have a head and body length of 290–390 mm and weigh 800–1100 g, while males have a head and body length of 300–550 mm and weigh 1000–2500 g. The animals have long silky grey fur above, which is paler below. The 200–290 mm long tail is black, with a prominent white crest along its rear half. The feet are white, and the ears are large and rabbit-like.

Distribution and habitat: Bilbies once lived over about two thirds of Australia, from the west coast north to Broome and south to Margaret River, but were apparently absent from the Swan Coastal Plain, and wetter parts of the jarrah and karri forests with heavy or stony soils. They extended east to western Queensland, to western and central New South Wales and north-western Victoria. They are now restricted to scattered populations in the northern parts of WA (Gibson and Great Sandy deserts, eastern Pilbara and the southern edge of the Kimberley). They also remain in parts of the central Northern Territory and south-west Queensland. Bilbies were once common,

but are now rare and declining. They lived in a wide variety of habitats, from open forest and woodland to desert loamy sands. In the south-west of WA they declined rapidly from about 1920, after the arrival of the fox. The last reports were from near Boyup Brook in 1975.

Biology and ecology: Bilbies spend the day in a deep burrow up to 3 m long and more than 2 m deep. Its entrance is often against a termite mound, spinifex hummock or shrub. Each animal has several burrows within its home range. After dark they leave the burrow to hunt and dig for food, which includes insects and their larvae, seeds, bulbs, fruit and fungi. Bilbies breed throughout the year, depending on seasonal conditions. Populations may move long distances to a new site, presumably because their current habitat becomes unsuitable in some way, for example, by food becoming scarce.

Threats: Bilbies are able to survive in the presence of feral cats, but an abundance of foxes, supported by rabbit populations, leads to their demise. Inappropriate fire regimes have probably caused their decline in some areas. Clearing would have contributed to their decline in the south-west, but was not the primary cause of local extinction.

Status: Vulnerable. Under *Western Shield*, bilbies have been reintroduced to Dryandra Woodland in the Wheatbelt and introduced to Francois Peron National Park in Shark Bay. At Dryandra, 45 bilbies were released between May 2000 and October 2001, and a small wild population has established. At Peron, 55 bilbies were released between 2000 and 2002 and, while a wild population has established, the project needs further monitoring before it can be properly evaluated.

Ngwayir, western ringtail possum *Pseudocheirus occidentalis*

Other names: Ngwayir, ngoor, ngoolangit, womp, woder (Nyoongar).

Description: Ngwayir have a head and body length of 300–400 mm, a tail 300–400 mm long and weigh 900–1100 g. They are dark brown (sometimes dark grey) above, and cream or grey to white below. The greyish-brown prehensile (grasping) tail has very short fur and a white tip of varying length. This species is distinguished from the common brushtail possum by its longer, non-bushy tail.

Distribution and habitat: The ngwayir is restricted to the south-west of WA. It once occurred from near Geraldton to the southern edge of the Nullarbor Plain, and inland to the eastern Goldfields. It has disappeared from most of its range, remaining in small parts of the jarrah forest (such as Perup and Kingston forests near Manjimup) and in coastal forests and woodlands, especially of peppermint (*Agonis flexuosa*), from Busselton to Albany.

Biology and ecology: Ngwayir spend most of their lives in trees and rarely come to the ground unless the canopy is too open. They eat leaves, especially those of peppermint, and shelter in tree hollows or in leaf nests in the canopy known as dreys. Births are mostly in winter, but they may breed throughout the year.

Threats: Most of their decline can be attributed to predation by foxes and, possibly, feral cats—ngwayir survive best in habitats with a dense canopy, where they don't need to come to the ground. Habitat destruction, especially for housing, is of concern, particularly between Busselton and Augusta. Recent research at Kingston Forest has shown that ngwayir were detrimentally affected by timber harvesting.

Status: Vulnerable. The ngwayir's conservation is guided by an interim recovery plan. Major issues include continued habitat destruction, mainly for housing, between Busselton and Margaret River, and appropriate management of forests. Ngwayir have been reintroduced to Leschenault Conservation Park and Yalgorup National Park.

Above Ngwayir are readily distinguished from brushtail possums by their long, short-furred tail.

Photo – Babs and Bert Wells /CALM

Above The ngilkait is Australia's most endangered mammal.

Photo – Jiri Lochman

Ngilkait, Gilbert's potoroo *Potorous gilbertii*

Other names: Gilbert's potoroo (English); ngilkait is the Nyoongar name for this animal.

Description: These dark grey, compact rat-kangaroos have bulging eyes. Males average about 930 g and females about 875 g.

Distribution and habitat: The ngilkait has been recorded alive only from the Albany area, however, subfossil records extend its former range westward to near Margaret River. For many decades the species was thought to be extinct, as the last specimens were taken between 1874 and 1879. The ngilkait was rediscovered by Elizabeth Sinclair and Adrian Wayne at Two Peoples Bay Nature Reserve in 1994. Searches in other south coast areas have so far been unsuccessful.

Biology and ecology: At Two Peoples Bay, ngilkait inhabit dense low heath on the slopes of Mt Gardner, where underground fungi comprise almost all the diet. The single young may be born at any time of the year. Females appear to be very choosy when selecting mates. Captive breeding has not been very successful—joeys have been born infrequently, with only eight young produced in the first six years of breeding. Captive ngilkait suffer from kidney failure due to calcium oxalate deposition. Some have died from this condition, which is thought to be due to a combination of genetic disease due to inbreeding in the small wild population and, possibly, also the artificial diet, which may not mimic closely enough the wild diet of truffle-like fungi.

Threats: The ngilkait's decline has been attributed to changed fire regimes and clearing, combined with predation by foxes and cats. At Two Peoples Bay, where foxes are controlled as part of *Western Shield*, the major threats are predation by feral cats and extensive fire. Dieback disease caused by *Phytophthora cinnamomi* degrades their habitat, and potoroos do not occupy infected areas, presumably because of the lack of food and/or shelter.

Status: Critically Endangered. Conservation is guided by a recovery team implementing a recovery plan. The ngilkait is the most endangered mammal in Australia, with well under 50 animals (possibly less than 25) remaining in a single wild population, and with captive breeding yet to prove successful. Artificial support for captive breeding (artificial insemination and cross-fostering of joeys) is being attempted, and searches are being conducted along the south coast in an attempt to locate additional populations.

Shark Bay boodie
Bettongia lesueur lesueur

Other names: Burrowing bettong (Shark Bay islands).

Description: Boodies from the Shark Bay islands have a head and body length averaging 350 mm and a tail of around 300 mm. The fur is grey to greyish-brown above and light grey below. When food has been plentiful, the tail is thick and fatty.

Distribution and habitat: Shark Bay boodies occur only on Bernier and Dorre islands in Shark Bay. Like their mainland relatives, they live in burrows, but these are sometimes solitary rather than always occurring in warrens.

Biology and ecology: Much of the food is obtained by digging. It probably includes fungi, tubers, bulbs, seeds and some animal material. A single young is born 21 days after conception, and remains in the pouch for 115 days. Maturity is reached at five months and breeding occurs throughout the year.

Threats: The introduction of predators or extensive or frequent fire on the islands would be potentially devastating.

Status: Vulnerable. Shark Bay boodies have been introduced to Heirisson Prong in Shark Bay, and to Roxby Downs in South Australia. Both areas have fox and cat-proof fences. Shark Bay boodies from Heirisson Prong were translocated to Faure Island in Shark Bay in 2002. Shark Bay boodies are being bred in large enclosures as part of the 'Return to Dryandra' project in the Wheatbelt, and there are plans to reintroduce them to the wild.

Above A Shark Bay boodie.

Photo – Babs and Bert Wells /CALM

Below A Barrow Island boodie.

Photo – Russell Lagdon/ ChevronTexaco Australia

Bottom Barrow Island boodies dig their warrens under caprock.

Photo – Andrew Burbidge /CALM

Barrow Island boodie   *Bettongia lesueur* **undescribed subspecies**

Description: Barrow Island boodies are smaller than mainland or Shark Bay boodies, with a head and body length of about 280 mm and a weight of 650–870 g. The fur is grey. Research by Ken Aplin at the WA Museum suggests that the Barrow Island boodie may be a different subspecies from the Shark Bay boodie.

Distribution and habitat: The Barrow Island boodie occurred on Barrow Island and the nearby small Boodie Island. Already very rare on Boodie Island before a black rat eradication project there in 1985, it disappeared soon afterwards, possibly because boodies were able to access poison grain dragged from the bait stations by the rats. It was then restricted to Barrow Island, where it utilised all habitats. Boodies are now abundant on Boodie Island once again, after being reintroduced there in 1993.

Biology and ecology: Barrow Island boodies dig warrens, usually on top of low limestone hills, often under caprock. They may move some kilometres from the warren at night to feed. On Boodie Island, some warrens are in sand among beach spinifex (*Spinifex longifolius*). One warren on Barrow Island has over 120 entrances, and limited counts suggest that warrens are occupied by at least half as many individuals as entrances. Barrow Island boodies feed on tubers, bulbs, seeds, fruits, fungi and termites. A single young may be born at any time of year, depending on seasonal conditions. Up to three joeys may be born in the same year in good seasons.

Threats: Introduction of predators such as cats, foxes and black rats.

Status: Vulnerable. A 1992 translocation to the Gibson Desert Nature Reserve failed due to feral cat predation. Mammals on Barrow Island are being monitored by a Department of Conservation and Land Management and ChevronTexaco study.

Above Barrow Island
spectacled hare-wallabies are
well adapted to a harsh
desert environment.

Photo – Babs and Bert Wells
/CALM

Barrow Island spectacled hare-wallaby
Lagorchestes conspicillatus conspicillatus

Description: This subspecies has a head
and body length of 400–470 mm, a
370–490 mm long tail, and weighs
1.6–4.5 kg (mean 3 kg). The name is
derived from a ring of rufous fur around
the eyes. This species has brown, white-
tipped fur above, with a white hip stripe,
and is off-white below.

Distribution and habitat: Barrow
Island spectacled hare-wallabies are now
restricted to Barrow Island. They once
occupied Hermite and Trimouille islands
in the Montebello group, but disappeared
due to feral cat (and possibly black rat)
predation. They inhabit spinifex (*Triodia
angusta, T. wiseana* and *T. pungens*)
hummock grassland.

Biology and ecology: During the day,
Barrow Island spectacled hare-wallabies
shelter inside large spinifex hummocks
(where the temperature rarely rises above
30°C). They emerge at late dusk to graze
on spinifex and browse herbs and shrubs.
Several hides are constructed within a
home range of about 8–10 ha. Research
has shown that they are well adapted to
an arid environment and have the lowest
water requirement of any marsupial.

The single young may be born at any
time of year, although there are peaks in
autumn and spring.

Threats: Threats include the potential
introduction of predators such as foxes,
cats and black rats to Barrow Island, and
the introduction of exotic diseases.
Extensive or frequent fire is not a major
threat on this large island, which has had
very large fires in the past.

Status: Vulnerable. 'Montebello
Renewal', a *Western Shield* project, is
eradicating feral cats and rats from the
Montebello Islands. Once this is achieved,
Barrow Island hare-wallabies can be
reintroduced. Mammals on Barrow Island
are being monitored by a joint
Department of Conservation and Land
Management and ChevronTexaco study.

On the mainland, another subspecies
(*Lagorchestes conspicillatus leichardti*) is
still found in the Pilbara and Kimberley,
and in the Northern Territory and
Queensland. It has disappeared from
more arid parts of its range, but has not
declined so much that it is considered
threatened.

Rufous hare-wallaby (Bernier Island)
Rufous hare-wallaby (Dorre Island)
Lagorchestes hirsutus bernieri
Lagorchestes hirsutus dorre

Description: Shark Bay rufous hare-wallabies are similar to mala (see p. 46), but are brown, rather than rufous, and slightly larger. It is unclear whether the two Shark Bay island populations of the rufous hare-wallaby are separate subspecies. Morphological studies have shown some differences, but genetic research has suggested that differences are minor.

Distribution and habitat: Restricted to Bernier and Dorre islands in Shark Bay, the hare-wallabies inhabit low scrub and spinifex on sandy soils.

Biology and ecology: During the day, hare-wallabies live in a 'squat', a shallow scrape or sometimes a shallow trench under a low shrub or spinifex hummock.

At night they feed on grass, including spinifex (*Triodia*), herbs and shrubs. A single young may be born at any time of the year and is carried in the pouch for about 15 weeks.

Threats: Introduction of predators or disease and extensive or frequent fire are the main threats.

Status: Vulnerable. Under 'Project Eden', part of *Western Shield*, hare-wallabies from the Shark Bay islands are being bred in captivity at Denham. In 2002, some were released into the wild on Peron Peninsula (where foxes had been eliminated and cats controlled), but feral cat predation was significant and the remaining animals were returned to captivity.

Banded hare-wallaby (Shark Bay islands)
Lagostrophus fasciatus fasciatus

Description: Although they look similar to other hare-wallabies (*Lagorchestes*), banded hare-wallabies are the only surviving species of a separate subfamily of macropod marsupials, which is more common in the fossil record. The head and body length is 400–450 mm, the tail length is 350–400 mm and the weight is between 1.3 and 2.1 kg. The upper fur is a dark grizzled grey, with transverse dark bands across the lower back and rump.

Distribution and habitat: This subspecies of the banded hare-wallaby is restricted to Bernier and Dorre islands in Shark Bay, where it lives in shrublands. The mainland subspecies is extinct.

Biology and ecology: The animals spend the day in communal groups sheltering under dense shrubs, and feed at night in open areas with scattered shelter. Grasses make up less than half of

their food. Leguminous shrubs and other plants form the rest of their diet. A single young is born. Births peak in late summer. Joeys spend about six months in the pouch and are weaned three months later.

Threats: Introduction of predators or disease, and extensive or frequent fire.

Status: Vulnerable. Under 'Project Eden', part of *Western Shield*, banded hare-wallabies from the Shark Bay islands are being captive-bred at Denham, along with rufous hare-wallabies. Some were released into the wild on Peron Peninsula in 2002, but, as with the rufous hare-wallabies, feral cat predation was significant and the remaining animals were returned to captivity.

Above Recherche rock-wallabies occur only on three islands off the south-west coast of WA.

Photo – Jiri Lochman

Below Research on remnant warru populations in the WA Wheatbelt first demonstrated that fox control can lead to a resurgence in mammal numbers.

Photo – Babs and Bert Wells /CALM

Recherche rock-wallaby *Petrogale lateralis hackettii*

Description: Similar to the warru (see below).

Distribution and habitat: Recherche rock-wallabies are restricted to Mondrain, Westall and Wilson islands in the Archipelago of the Recherche.

Biology and ecology: Recherche rock-wallabies shelter in rock outcrops, but forage over most of the islands. Their biology and ecology has not been studied, but is probably similar to that of warru (see below).

Threats: Introduction of predators or disease and extensive or frequent fire are the main threats.

Status: Vulnerable. All islands in the Archipelago of the Recherche are nature reserves.

Warru, black-flanked rock-wallaby *Petrogale lateralis lateralis*

Other names: Black-footed rock-wallaby (English), bokal, moororong (Nyoongar), kakuya, lungkarrpa, pakultarra, rukapiki, tanpa, tjinangalku, tjirti, warru, wartilara, wokartji (western desert Aboriginal dialects).

Description: Warru have a head and body length of 500–545 mm and weigh 2.1–4.5 kg. They are dark reddish-brown above, with a grey nape and shoulders. There is a distinct black eye stripe and a white cheek stripe, and white at the base of the ears. A white stripe, with a dark stripe just above it, runs along the side of the body. The long, 520–580 mm tail ends in a black brush.

Distribution and habitat: Warru were once very widespread but scattered in the Great Sandy, Little Sandy, Gibson and northern Great Victoria deserts, the Central Ranges region, Ashburton, North West Cape and the south-west from Kalbarri to the southern Wheatbelt. Their distribution is now greatly reduced, with remnant populations in the Wheatbelt near Kellerberrin, Cape Range, the southern edge of the Pilbara, and a very small colony in the Calvert Range (south of Lake Disappointment). This subspecies also occurs on Barrow and Salisbury islands, but disappeared from Depuch Island (near Whim Creek) when foxes waded across from the mainland.

Biology and ecology: Warru shelter in rock piles and screes during the day, emerging at night to feed on grasses and shrubs. Year-round breeding is possible when conditions are favourable. A single joey is born, usually in spring. Pioneering research on warru by Jack Kinnear and colleagues near Kellerberrin first demonstrated that fox control could lead to a resurgence in numbers of critical weight range mammals (see p. 33).

Threats: Fox predation, large wildfires and, possibly, competition from rabbits are the main threats. Fire is a threat to isolated desert populations.

Status: Vulnerable. All known colonies in the south-west are protected by fox control under *Western Shield*, as is that of the Calvert Range.

Black-footed rock-wallaby (MacDonnell Ranges)
Petrogale lateralis undescribed subspecies

Other names: Warru (western desert Aboriginal dialects).

Description: This subspecies has a head and body length of 450–520 mm, a tail 507–597 mm long, and weighs 2.8–4 kg. Though it is similar in colour to the warru (see p. 66), its fur is shorter and less dense.

Distribution and habitat: This rock-wallaby once inhabited rocky ranges and granite boulder piles in the central ranges of the Northern Territory, South Australia and WA. It has disappeared from all granite tors and boulder piles, and remnant colonies persist only in the most rugged pockets of rock outcrop in major ranges. In WA, it is now known from Walter James Range, Bell Rock Range, Cavenagh Range, Rawlinson Range and Townsend Ridges (south of Warburton).

Biology and ecology: Similar to the warru.

Threats: The subspecies has declined since the fox arrived in central Australia, and this decline is continuing. Fox baiting is being implemented at the Townsend Ridges in conjunction with the Ngaanyatjarra Council, who are also searching for other remnant populations in their region.

Status: Vulnerable.

Above A black-footed rock-wallaby of the MacDonnell Ranges race. Rock-wallabies have a remarkable ability to scale almost-vertical cliffs.

Photo – Don Langford /Bluesky Enterprises

Black-footed rock-wallaby (West Kimberley)
Petrogale lateralis undescribed subspecies

Description: Similar to the warru (see p. 66).

Distribution and habitat: This undescribed subspecies occurs in the Erskine Range/Done Hill area, Grant Range and Edgar Range in the south-west Kimberley. The rock-wallabies live in high sandstone cliffs and screes, with hummock grassland and occasional figs and low shrubs.

Biology and ecology: Similar to the warru.

Threats: Foxes are absent or uncommon in this area, but would be a significant threat should they establish. Inappropriate fire regimes are a current threat. Cattle grazing in feeding areas around the base of outcrops may be a threat.

Status: Vulnerable. The Department of Conservation and Land Management has conducted searches of the area with Macquarie University to clarify the genetic status of this taxon. Further populations were located in the Dogspike Hill area. The Erskine Range population has been monitored occasionally, and the rock-wallabies were observed to persist in good numbers after a large fire in the area.

Above Barrow Island euros are small enough to shelter in the shade of large spinifex hummocks.

Photo – Andrew Burbidge /CALM

Barrow Island euro *Macropus robustus isabellinus*

Description: The Barrow Island euro is a small euro, with little difference in size or colouration between males and females. Males weigh 10–15 kg and females weigh 7.5–9.5 kg. Mainland euros are much larger, with males being redder in colour and considerably larger (up to 45 kg) than females (averaging about 20 kg).

Distribution and habitat: This subspecies occurs only on Barrow Island, where it shelters in caves and mangroves and feeds in spinifex hummock grassland. Nowadays, it also shelters in the shade cast by oilfield facilities.

Biology and ecology: Barrow Island euros graze on spinifex and browse on shrubs. A single joey can be born at any time of the year, depending on seasonal conditions.

Threats: Threats include the introduction of predators or disease. Some euros have been killed by vehicles, but road kills are now infrequent as animals have learned to avoid vehicles. Most oil pumps on Barrow Island are timed to turn on infrequently, and animals sheltering in the shade were being killed by the pump counter-weights. ChevronTexaco installed matting with sharp cones on it under the oil pumps and this has greatly reduced deaths from this source.

Status: Vulnerable. While Barrow Island is a nature reserve, it is also an oil production lease and a major industrial development is proposed for the island.

Quokka   *Setonix brachyurus*

Other names: Short-tailed wallaby (English), bangop, kwoka (Nyoongar).

Description: Quokkas have a head and body length of 400–540 mm and a 245–310 mm tail. They are a grizzled greyish-brown above, slightly paler below, and have long, thick, coarse fur.

Distribution and habitat: Quokkas are restricted to the south-west of WA, from near Gingin to the Green Range east of Albany. They also occur on Rottnest and Bald islands. On the mainland they once lived in a wide variety of habitats, especially swamps and other thick vegetation including coastal thickets. Their mainland range is now greatly reduced and fragmented. The species now lives only in the densest vegetation, often on the edge of swamps in forested areas, from near Dwellingup to near the south coast, and in near-coastal thickets east to Green Range east of Albany. In the northern jarrah forest, they require a complex mosaic of thicket, some recently burnt and some long-unburnt.

Biology and ecology: Quokkas have been much studied on Rottnest Island and their biology is well understood. On Rottnest, they breed only once a year, in late summer, with joeys leaving the pouch in August and becoming independent in October. On the mainland, breeding can occur throughout the year.

Threats: Island populations appear secure. On the mainland, threats include predation by foxes and habitat degradation through inappropriate fire. Quokkas are continuing to decline in the northern jarrah forest. The effects of logging adjacent to their habitat has not been studied, however, quokkas have survived more than 100 years of logging.

Status: Vulnerable. Most mainland populations are now protected by fox control through *Western Shield*, however, populations in the northern jarrah forest have not recovered and some have disappeared. Some mainland populations are being monitored.

Below Mainland quokkas inhabit dense thickets and are hard to observe.

Photo – Bill Belson /Lochman Transparencies

Orange leaf-nosed bat *Rhinonicteris aurantius*

Other names: Orange horseshoe-bat.

Description: Orange leaf-nosed bats have a head and body length of 45–53 mm and a tail 24–28 mm long. They weigh 8–10 g. The orange fur is darker around the eyes, and the ears are small and acutely pointed.

Distribution and habitat: This species is found in the Kimberley and Pilbara, where it roosts in deep, warm, moist caves and old mine adits (an adit is an almost horizontal entrance to a mine).

Biology and ecology: Orange leaf-nosed bats emerge from their roosts at dusk to feed on moths and other flying insects. They are very sensitive to human interference. They quickly take flight if disturbed, and may abandon disturbed roosts completely.

Threats: Roost site disturbance and adit collapse are significant threats in the Pilbara. Frequent and extensive fire may reduce their food supply.

Status: Vulnerable. The Pilbara population is very small, with few known roost sites. Many of these are in old mine adits that are likely to collapse in the future. Considerable resources have been put into searches for roost sites.

Palyoora, plains rat *Pseudomys australis*

Above Palyoora have not been recorded in Western Australia since 1969.

Photo – Babs and Bert Wells /CALM

Description: The palyoora has a head and body length of 100–140 mm, an 80–120 mm long tail and weighs 50–80 g. It is grey to greyish-brown above, white or cream below, and the tail is lighter towards the tip. The ears are relatively large.

Distribution and habitat: The species was once widespread in south-eastern WA including the eastern Goldfields, mallee-dominated vegetation in the south-eastern Wheatbelt, eastern south coast, and Nullarbor and Roe plains in WA and parts of South Australia, Victoria and Queensland. In WA, the last record was from near Mundrabilla in 1969. Palyoora are more common in north-eastern South Australia.

Biology and ecology: Palyoora construct shallow, complex burrow systems, often grouped about 10 m apart to form colonies interconnected by surface runways. In good seasons these may spread over large areas. Palyoora are thought to feed mainly on seeds, and are well-adapted to living in arid areas. Breeding occurs after heavy rain. Litters may contain up to seven young, although three to four is more usual.

Threats: Predation by foxes and feral cats, habitat degradation by hard-hooved stock and rabbits.

Status: Vulnerable. The species is probably extinct in WA, but still occurs in South Australia.

Djoongari, Shark Bay mouse *Pseudomys fieldi*

Other names: Alice Springs mouse (English), konding (Nyoongar). The Shark Bay mouse (originally named *P. praeconis*) is now considered to be the same species as the Alice Springs mouse (*P. fieldi*). The latter scientific name has priority.

Description: Djoongari have a head and body length of 85–100 mm, a tail 115–125 mm long and weigh 30–50 g. They are a grizzled dark brown above and white below. The furred tail is grey above and white below.

Distribution and habitat: The species once occurred throughout much of the western two thirds of Australia south of the tropics. It became extinct on the mainland soon after European settlement, and for many years was restricted to Bernier Island in Shark Bay until translocations began in the 1990s. On Bernier Island, djoongari mainly inhabit coastal dunes dominated by beach spinifex (*Spinifex longifolius*) and coastal daisy (*Olearia axillaris*), but recent surveys suggest that the species is found in most coastal sandy areas around the island and also occurs at lower densities in inland *Triodia/Acacia* heath. Nothing is known of its preferred habitat on the mainland, though it is likely to have inhabited deep sandy soils supporting *Spinifex* and *Triodia* species.

Biology and ecology: Djoongari feed on plant material and invertebrates such as insects and spiders. They do not appear to use burrows as much as most other *Pseudomys* species. They build tunnels and runways in heaps of seagrass piled up on Bernier Island beaches during winter storms, and use above-ground nests as daytime refuges. More use of burrows is made during the breeding season. On Bernier Island, djoongari breed at any time between May and November, with sub-adults (less than 30 g body weight) entering the population between November and March. Litter sizes of up to five have been observed. Animals on Bernier Island live for at least two years. They take about 100 days to reach adult size.

Threats: Their extinction on the mainland was probably due to feral cats, although trampling of burrow systems by stock may have contributed in some areas. Threats to the island populations are the introduction of predators and disease, and inappropriate fire regimes.

Status: Vulnerable. Conservation is being guided by a recovery plan and recovery team. Djoongari have been successfully introduced to Doole Island (Exmouth Gulf) and North West Island (Montebello Islands). A reintroduction to Heirisson Prong in Shark Bay was unsuccessful, possibly because of predation by sand goannas (*Varanus gouldii*).

Above Barrow Island mouse.

Photo – Kathie Atkinson

Barrow Island mouse
Pseudomys nanus ferculinus

Description: The Barrow Island mouse has a head and body length of 30–34 mm, and weighs 35–55 g. It is fawn with longer dark brown hairs above, and pronounced lighter coloured eye rings. The tail is dark above and light below.

Distribution and habitat: This subspecies is restricted to Barrow Island, where it occupies all vegetated habitats.

Biology and ecology: Barrow Island mice live in shallow burrows under acacia litter or among beach spinifex (*Spinifex longifolius*) in sandy habitats. On rocky habitats they occupy dense *Triodia* hummocks. Breeding appears to follow heavy rain.

Threats: Introduction of predators or disease.

Status: Vulnerable. The species is being monitored, with other mammals on Barrow Island, by a joint Department of Conservation and Land Management and ChevronTexaco study.

Dayang, heath mouse   *Pseudomys shortridgei*

Above Two widely-separated populations of dayang occur in the south-west of WA and western Victoria.

Photo – Babs and Bert Wells /CALM

Other names: Heath rat.

Description: Dayang have a head and body length of 33–40 mm and a tail of a similar length. They weigh 35–70 g, with a pes (foot) length of 19–22 mm. They are greyish-brown above, flecked with black, and paler below. The face is blunt (similar to a bush rat), and the tail is hairy. Dayang can be distinguished from mootit (southern bush rats) by the lack of rings on their tail.

Distribution and habitat: The two widely separated populations of dayang— in the south-west of WA and western Victoria—appear to be closely related. In WA, dayang inhabit kwongan (heath) vegetation and arid shrublands. They occur at Lake Magenta and Dragon Rocks nature reserves, Fitzgerald River National Park and the Ravensthorpe Range.

Biology and ecology: In WA, dayang seem to prefer vegetation that has not been burnt for more than 15 years.

Threats: Predation by cats and foxes, and inappropriate fire regimes, threaten the species.

Status: Vulnerable. As studies of dayang in the south-west have been limited, more are needed to clarify their conservation requirements. Limited taxonomic research suggests that eastern and western populations may be different subspecies. Further genetic studies are underway.

Wopilkara, greater stick-nest rat   *Leporillus conditor*

Other names: House-building rat (English). For Aboriginal names see djooyalpi (p. 44).

Description: Wopilkara have a head and body length of 170–260 mm and weigh 180–450 g. They are grey above and creamy white below. Their ears are long, their eyes are large and the snout is rather blunt. The 145–180 mm long tail is evenly furred.

Distribution and habitat: The species was once found through much of semi-arid and arid southern Australia, from the central west coast of WA to western New South Wales and north-western Victoria. In WA, it occurred in the Carnarvon Basin, Geraldton sandplains, the Murchison and Yalgoo region, the eastern Goldfields, the Nullarbor and on Roe Plain. They had a more southerly range than the djooyalpi, but the distribution of the two species overlapped. Wopilkara had become rare by the mid-nineteenth century and were extinct on the mainland by the 1930s. They survived only on Franklin Island in South Australia. Captive-bred animals from this population have now been established on Salutation Island and on Heirisson Prong, both in Shark Bay, and on some South Australian islands and mainland sites.

Biology and ecology: Wopilkara are exclusively herbivorous, eating leaves and fruits of succulent plants. Breeding is in autumn and winter. One or two well-developed young are born after a gestation period of more than six weeks. Many early explorers were impressed by the large stick and stone nests—up to 1 m high and 1.5 m in diameter—constructed by this animal and by djooyalpi (lesser stick-nest rats). A nest of grass is constructed in the centre and houses communities of up to 20 animals.

Threats: Predation by foxes and feral cats.

Status: Vulnerable.

Central rock-rat *Zyzomys pedunculatus*

Other names: None. Antina was proposed as the Australian name for this species, but recent information suggests that the word was not used by Aboriginal people for this animal.

Description: Central rock-rats have a head and body length of 110–125 mm, and a tail length of 110–140 mm, and weigh 70–100 g. Adults are strongly built and have large ears and prominent eyes. The long, harsh fur is yellowish-brown above and cream below. The well-furred, pale brown tail is thicker at the base and tapers towards the tip.

Distribution and habitat: Despite several searches, central rock-rats were not collected between 1960 and 1996, when some were caught in the MacDonnell Ranges, Northern Territory. A captive population was established at Alice Springs and later some were transferred to Perth Zoo. The species has never been found alive in WA, but its former distribution has been inferred from subfossil deposits. In WA, it once occurred from North West Cape, through the Gascoyne and Murchison to the Little Sandy and Great Sandy deserts. No subfossil material has been found in the Pilbara, though this seems a likely area. Central rock-rats inhabit arid scree slopes and rock piles.

Biology and ecology: Central rock-rats are mainly herbivorous, eating seeds and leaves. A few insects may also be eaten.

Threats: They are threatened by predation by cats, black rats and foxes.

Status: Critically Endangered. The Northern Territory Department of Infrastructure, Planning and Environment has developed a conservation program and is captive-breeding central rock-rats at the Desert Wildlife Park in Alice Springs. A small colony is now established at Perth Zoo. The WA Department of Conservation and Land Management prepared an interim recovery plan in 1997, but this covered actions only in WA, the most important of which was to search for populations. A search by departmental scientific staff in Cape Range, where subfossil remains are common in caves, did not locate any animals. Black rats were trapped in habitats once occupied by central rock-rats, and it seems likely that they have contributed to the rock-rat's decline. Introduction to one of the Montebello Islands may be possible once it is confirmed that black rats have been eradicated there. Further searches are being carried out during the Pilbara Region Biological Survey.

Blue whale ('true' subspecies)
Balaenoptera musculus musculus

Description: The largest animals ever, blue whales grow to 30.5 m, weigh up to 150 tonnes or more, and may live for up to 80–90 years. There are two subspecies: the 'true' blue and the so-called 'pygmy' blue (*Balaenoptera musculus brevicauda*), which is still very large at 21–24 m long. The long, slender, laterally-compressed, mottled bluish-grey body distinguishes blue whales from other species.

Distribution and habitat: This oceanic species is found worldwide, and migrates between warm-water breeding areas and cold-water feeding areas. Most blue whales seen near the WA coast are the pygmy subspecies, which occurs only in the southern hemisphere and is most abundant in the Indian Ocean.

Biology and ecology: Southern hemisphere 'true' blue whales feed almost exclusively on Antarctic krill.

Threats: Blue whales were reduced drastically by whaling. The southern hemisphere population was estimated at about 160,000–240,000 prior to whaling, including 10,000 pygmy blues. The 'true' blue population is now estimated to be less than 1000, plus about 6000 pygmy blues. They are still threatened by illegal whaling by fleets primarily targeting other species, by seismic and defence operations, collision with ships, entanglement in fishing gear and pollution causing toxic chemicals to accumulate in body tissues.

Status: Endangered. The subspecies has not recovered significantly since protection.

Above The world's largest animal, the blue whale has been greatly reduced in numbers by commercial whaling.

Photo – Doug Coughran /Lochman Transparencies

Sei whale   *Balaenoptera borealis*

Description: Males are about 17 m long. Females are about 21 m and weigh up to 26 tonnes. This long streamlined whale has a sharp-pointed but flat-topped rostrum (front of the upper head) with a single central ridge. The whale's upper surface is grey with occasional white spots. Its throat is white and its belly is grey. Bryde's whales have three ridges along the rostrum. Fin whales have a bi-coloured head and a larger dorsal fin than comparable whales.

Distribution and habitat: This worldwide, oceanic species undertakes long migrations between warm water breeding grounds and colder water feeding grounds, however, in the southern hemisphere it does not migrate as far south as some other baleen whales. It does not come close to shore and is infrequently recorded in Australian waters.

Biology and ecology: Sei whales feed mainly on copepod crustaceans and occasionally on krill. Sexual maturity is reached in 7–11 years and at 13–14 m.

Threats: Numbers were greatly reduced over a short period of time (from about 1960 to 1977) by whaling. Numbers in the southern hemisphere were probably about 100,000 before whaling began, but by 1977 there were probably only about 25,000 left.

Status: Vulnerable. All whales are protected within Australian waters (up to 200 nautical miles offshore). Beyond this, commercial whaling is subject to International Whaling Commission decisions. Some nations, such as Japan, continue to hunt whales in the southern oceans. Whale conservation is primarily a Commonwealth Government concern.

Humpback whale *Megaptera novaeangliae*

Description: Humpback whales grow to 13–14 m and can weigh up to 45 tonnes. They have a stout body and a knobby head. They are easily distinguished from other whales by their very long flippers, which may be a third of their body length. The dorsal fin, nearly two thirds of the way along the back, is reduced to a fleshy hump.

Distribution and habitat: Humpback whales have a worldwide, coastal distribution. They migrate annually between warm-water breeding and cold-water feeding grounds. The population found off the WA coastline is genetically distinct from other populations, including the one that migrates along Australia's east coast. Northward-migrating animals may be seen off the south coast, but swim too far offshore along the west coast to be frequently sighted from shore or from small boats. Calving occurs off the Kimberley. When heading south, in spring and early summer, most animals swim within ten nautical miles of the shore.

Biology and ecology: Sexual maturity is attained when the whale is about 11.5 m, from four to ten years of age. Calving takes place from June to October and females calve every three years. Humpback whales feed almost exclusively on Antarctic krill.

Threats: Numbers of humpback whales were greatly reduced by whaling and onshore whaling stations on the WA mainland contributed to the decline of the species. It is thought that the population was reduced to only 5–6% of its original size by 1963, when this species became protected. Continued illegal catches by high-seas, foreign-owned whaling fleets delayed recovery until the mid-1970s. Humpback numbers are now increasing, with numbers of the population that migrates along the WA coast (the largest population in the world) currently exceeding 20,000. Current threats include entanglement in fishing gear, and disturbance by the whale-watching industry and other boats.

Status: Vulnerable. The Department of Conservation and Land Management manages the whale-watching industry and seeks, via education, to minimise disturbance of migrating and breeding whales by boats.

Opposite Humpback whales, once threatened by whaling, are making a comeback and are now a tourist attraction.

Photo – Geoff Taylor /Lochman Transparencies

Above Southern right whales can be seen during winter, close to shore along WA's southern coastline.

Photo – Geoff Taylor /Lochman Transparencies

Below A fin whale in the Southern Ocean.

Photo – Dave Watts

Southern right whale *Eubalaena australis*

Description: Southern right whales can reach 17 m long and weigh up to 80 tonnes. They may live for 50 or more years. Their distinct, rotund body has no dorsal fin. Apart from striking horny growths ('callosities') around the mouth, above the eyes and on top of the head, these whales are mostly bluish-black or dark brown.

Distribution and habitat: The southern right whale is a southern hemisphere, circumpolar species that migrates from warmer northern latitudes to cold southern ones. It does not, however, migrate as far north as many other baleen whales, being limited to about 30°S.

Biology and ecology: The southern right whales live in the open ocean in summer, feeding on smaller plankton, mainly larval crustaceans and copepods. They calve between June and August. In WA, females come very close to the shore from about Perth (sometimes further north) southwards, so as to give birth and feed their calves. Greatest numbers are found along the south coast. Sexual maturity is reached at nine to ten years, when the animal is 12–13 m long, and a female usually calves every three years.

Threats: Right whales were considered the 'right' whales to hunt, especially in the early days of sailing ships and open boats. Whaling, particularly in the early 1800s, reduced numbers off Australia to a tiny remnant of the original population. Catching continued until the 1930s and, despite protection, some were killed until at least the late 1960s. The Australian southern right whale population is now recovering, with an increase of about 7% each year. Current threats include collision with ships, disturbance at calving grounds, seismic and defence operations, and entanglement in fishing gear.

Status: Vulnerable. CALM assists with research to establish population numbers and trends, and is working to prevent disturbance by the whale-watching industry and private boats. Marine park proposals cover some of the more important calving areas.

Fin whale *Balaenoptera physalus*

Description: These baleen whales grow to 25–27 m long, and weigh up to 90 tonnes. They have a narrow, elongated, laterally-compressed body and a tall dorsal fin. The upper parts are dark grey or dark brown, but the undersurfaces of the flukes and flippers have white areas. The head is bi-coloured, with a white right lower jaw and white baleen plates, a feature that distinguishes it from other whales.

Distribution and habitat: The fin whale is a fast-swimming oceanic species found worldwide. It migrates between warm-water breeding grounds and cold-water feeding grounds, but migration routes do not follow coastlines.

Biology and ecology: Fin whales feed largely on Antarctic krill. They reach sexual maturity at about six to ten years, when they are around 19 m long.

Threats: It is thought that the southern hemisphere population once numbered about 500,000. Now, there may only be about 25,000 left, as numbers have been reduced drastically by whaling.

Status: Vulnerable.

References

Abbott, I. (2001). Aboriginal names of mammal species in south-west Western Australia. *CALMScience* 3, 433–486.

Bannister, J.L., Kemper, C.M. and Warneke, R.M. (1996). *The action plan for Australian cetaceans.* Australian Nature Conservation Agency, Canberra.

Duncan, A., Baker, G.B. and Montgomery, N. (eds) (1999). *The action plan for Australian bats.* Environment Australia, Canberra.

Braithwaite, R.W., Morton, S.R., Burbidge, A.A. and Calaby, J.H. (1995). *Australian names for Australian rodents.* Australian Nature Conservation Agency in association with CSIRO Division of Wildlife and Ecology.

Burbidge, Andrew (1997). Antina (*Zyzomys pedunculatus*) Interim Recovery Plan 1996 to 1998. Interim Recovery Plan No. 5. In Interim Recovery Plans 4–16 for Western Australian critically endangered plants and animals, ed by Jill Pryde, Andrew Brown and Andrew Burbidge. Western Australian Wildlife Management Program No. 29.

Burbidge, Andrew A., Johnson, K.A., Fuller, Phillip J. and Southgate, R.I. (1988). Aboriginal knowledge of the mammals of the central deserts of Australia. *Australian Wildlife Research* 15, 9–39.

Burbidge, Andrew A. and McKenzie, N.L. (1989). Patterns in the modern decline of Western Australia's vertebrate fauna: causes and conservation implications. *Biological Conservation* 50, 143–198.

Churchill, S. (2001). Survey and Ecological Study of the Sandhill Dunnart, *Sminthopsis psammophila*, at Eyre Peninsula and the Great Victoria Desert. South Australian Dept. of Environment and Heritage/The Natural Heritage Trust, Adelaide, South Australia.

Churchill, S. (2001). Recovery plan for the sandhill dunnart (*Sminthopsis psammophila*). Department of Environment and Heritage, Adelaide.

Leake, B. (1962). *Eastern wheatbelt wildlife.* Author, Perth.

Lee, A.K. (1995). *The action plan for Australian rodents.* Australian Nature Conservation Agency, Canberra.

Mangglamarra, G., Burbidge, A.A. and Fuller, P.J. (1991). *Wunambal words for rainforest and other Kimberley plants and animals.* In Kimberley Rainforests of Australia, ed by N.L. McKenzie, R.B. Johnston and P.G. Kendrick. Surrey Beatty and Sons, Chipping Norton, New South Wales, pp. 413–21.

Maxwell, S., Burbidge, A.A. and Morris, K. (eds) (1996). *The 1996 Action Plan for Australian marsupials and monotremes.* Prepared by the Australasian Marsupial and Monotreme Specialist Group, IUCN Species Survival Commission. Wildlife Australia, Canberra.

Menkhorst, P. and Knight, F. (2001). *A field guide to the mammals of Australia.* Oxford University Press, Melbourne.

Morris, K., Speldewinde, P. and Orell. P. (2000). Djoongari (Shark Bay Mouse) recovery plan 1992–2001. Western Australian Wildlife Management Program No. 17. Department of Conservation and Land Management, Bentley.

Orell, P. and Morris, K. (1994). Chuditch recovery plan, 1992–2001. Wildlife Management Program No. 13. Department of Conservation and Land Management, Como.

Sattler, P. and Creighton, C. (2002). *Australian terrestrial biodiversity assessment 2002.* National Land and Water Resources Audit, Canberra.

Serena, M., Soderquist, T.R. and Morris, K. (1991). The chuditch (*Dasyurus geoffroii*). Wildlife Management Program No. 7. Department of Conservation and Land Management, Como.

Strahan, R. (ed) (1985). *The mammals of Australia.* Reed Books, Sydney.

Birds

chapter 5

Above A male Dirk Hartog Island southern emu-wren.

Opposite A male Gouldian finch enters a nest spout. Gouldian finches have declined because of changing fire regimes and cattle grazing.

Photos – Babs and Bert Wells /CALM

The many amateur and professional birdwatchers and ornithologists make birds the best known group of organisms in Australia. Birds Australia, the largest ornithological society, is a major non-government organisation dedicated to the study and conservation of birds—no organisation concerned with any other group of animals has such a large membership or lobbying power.

Because most birds can fly long distances in a short time, many species that do not normally reside in Western Australia may turn up in the State from time to time, making it difficult to provide an accurate figure for the number of bird species. Currently, it is considered that 472 resident or migratory species regularly visit WA, including State waters; the total number of species and subspecies is 522.

While land clearing and degradation have affected WA birds, they have not suffered from predation by invasive species to the same extent as mammals. Australia has far fewer extinct birds than extinct mammals. *The action plan for Australian birds 2000* lists eight species as extinct, plus one more that was a non-resident vagrant. Of these, only the paradise parrot (*Psephotus pulcherrimus*) was from mainland Australia. Two were from continental islands—species of dwarf emus from King and Kangaroo islands. The rest were from oceanic islands such as Lord Howe, Norfolk and Macquarie. A further 17 subspecies are listed as extinct. Again, most are from oceanic islands. Worldwide, most extinct birds are from oceanic islands such as Hawaii, New Zealand and Mauritius. Thus, three (2.3%) of the world's 129 extinct birds were from the Australian mainland and continental islands.

No bird species from WA has become extinct. While two subspecies have disappeared from WA, other subspecies of both these birds persist elsewhere in Australia. The proportion of species and subspecies that are extinct is 0.4%. The 2000 *IUCN Red List of threatened species* includes 1192 threatened bird species, and Australia has about 5.5% of the world total.

In 2003, 28 bird species were listed as threatened in WA. As well as the two subspecies of extinct birds, 14 subspecies were listed as threatened, making a total of two extinct and 42 threatened bird species and subspecies. Two are Critically Endangered—the Amsterdam albatross (*Diomedea amsterdamensis*) and night parrot (*Pezoporus occidentalis*)—eight are Endangered and 32 are Vulnerable. The proportion of WA species listed as threatened is 5.9%, while the proportion of species plus subspecies is 8%.

The main threats to terrestrial and wetland birds have been land clearing and land degradation due to grazing and changed fire regimes. Introduced predators such as the European fox and feral cat are a threat to ground-nesting species—such as nightjars, the malleefowl and the bush stone-curlew—though some ground-nesting or near ground-nesting birds—such as quail and the common bronzewing (which is common in the arid zone despite often nesting close to the ground)—appear to have been unaffected. The Critically Endangered night parrot probably declined to near extinction because of nest predation by cats and foxes, but changed fire regimes and stock and rabbit grazing may also have contributed. A suite of ground-dwelling birds found along the south coast (the noisy scrub-bird, western bristlebird, western whipbird and western ground parrot) are considered to have declined primarily because of an increase in the extent and frequency of fire, exacerbated by land clearing and, possibly, predation. Many Wheatbelt bird species are now very rare because of clearing. Fortunately, many of these have ranges that extend beyond cleared areas.

Birds that require tree hollows for nesting have suffered greatly from loss of hollows through land clearing and timber production. This loss is now being increasingly exacerbated by hollow invasion by feral honey bees (*Apis mellifera*), which use hollows as hives, preventing access by birds. Competition for food resources by honey bees may also be affecting native birds.

A significant proportion of listed birds are seabirds threatened by longline fishing conducted by high-seas tuna-fishing fleets. Thirteen species of albatrosses, the southern giant-petrel and the white-chinned petrel are listed because longline fishing gear drowns birds that take baited hooks. Other more common seabirds, such as shearwaters, are also being killed in large numbers, and may be listed in future if this threat does not abate.

For more information on WA's threatened birds, read the Bush Book *Threatened and rare birds of Western Australia*, by John Blyth and Allan Burbidge, published by the Department of Conservation and Land Management.

Left Flocks of the Endangered Carnaby's black-cockatoo are often seen in the Perth metropolitan area.

Photo – Babs and Bert Wells/CALM

Extinct subspecies

Lewin's rail   *Rallus pectoralis clelandi*

Other names: Lewin's water rail, slate-breasted rail. Seven other subspecies of Lewin's rail occur in eastern Australia, New Guinea, Indonesia and the Philippines.

Description: This bird was 21–28 cm long. It had a bright chestnut head and nape. Its wings and back were marbled or mottled in various shades of black, brown and buff, while the belly and flanks were barred black and white. The long, slender bill was bright pink at the base, with a grey tip. The WA subspecies had a clearly demarcated grey breast. It inhabited swamps with dense low emergent vegetation, especially those with some permanent water, in the south-west forests. It has been recorded from near Albany, Margaret River, Bridgetown and Manjimup.

Approximate date of extinction:
The last reliable report of Lewin's rail was from a swamp near Manjimup in 1932. Searches by several ornithologists, including Department of Conservation and Land Management staff, have failed to locate it since, but it is possible that it remains in dense remote swamps.

Probable cause of extinction: Not known. The Lewin's rail was probably rare at settlement. Increased fire size and intensity in the nineteenth and early twentieth centuries, as well as clearing, may have caused its extinction.

Above The Tasmanian subspecies of Lewin's rail (*Rallus pectoralis muelleri*).

Photo – T A White /NatureFocus

Below The closely-related eastern subspecies of the rufous bristlebird.

Photo – Hans and Judy Beste /Lochman Transparencies

Rufous bristlebird (western subspecies)   *Dasyornis broadbenti litoralis*

Description: Eastern rufous bristlebirds are 25–27 cm long and have a wingspan of 22.5 cm—limited data suggest WA birds may have been slightly smaller. WA rufous bristlebirds were medium-sized, sturdy, thrush-like and ground-dwelling, with short, rounded wings. They were mainly brown above with rich rufous on the top of the head and an off-white chin and throat. They also had a mottled dark brownish breast with white scalloping, an off-white belly, a long brown broad tail and brown under the tail coverts. Rufous bristlebirds occurred in dense coastal scrub between Cape Mentelle and Cape Naturaliste in the far south-west of the State.

Approximate date of extinction:
The western subspecies was discovered in 1901, and last recorded some time between 1906 and 1908. Most of its former habitat is now in the Leeuwin-Naturaliste National Park. Unconfirmed reports in 1940, 1977 and 1980 generate some hope that the subspecies may yet be rediscovered.

Probable cause of extinction: Too frequent burning of habitat in the early twentieth century, to promote feed for cattle, is considered to be the main cause of extinction.

Threatened species and subspecies

Malleefowl   *Leipoa ocellata*

Other names: Ngawoo (Nyoongar) (recorded as gnow by early Europeans, for example, as in Gnowangerup).

Description: This robust bird, 60–70 cm in length, has large feet and a grey head and breast. Its upper parts are grey, barred with white and brown, and the underparts are light fawn.

Distribution and habitat: Once very widespread and common in the southern half of WA, as well as in other states, malleefowl have disappeared from the Gibson Desert and most pastoral country. Some small remnant populations remain in the Great Victoria Desert. Malleefowl are still reasonably common north of the agricultural zone, from Shark Bay to near Yalgoo, and east of the central Wheatbelt agricultural country. Scattered populations remain in remnant Wheatbelt vegetation and near the south coast, including the Roe Plain to the south of the Nullarbor Plain.

Biology and ecology: These birds inhabit tall mallee, low woodland or acacia scrub on sandy soils, with a fairly complete canopy and abundant litter. Malleefowl build a large mound of sand and leaf litter in which the male incubates the eggs by controlling the temperature within the mound. They eat seeds, herbs and fruits, as well as ground-dwelling invertebrates.

Threats: Predation by foxes, especially of chicks, is a major threat. Land clearing has removed much habitat, and inappropriate fire regimes are a significant problem, as fires make habitat unsuitable for some years. After very large fires it may take decades for malleefowl to reinvade an area.

Status: Vulnerable. The Department of Conservation and Land Management is part of a national malleefowl recovery team. Conservation work in WA is coordinated through the Malleefowl Recovery Network. Fox control via *Western Shield* is helping malleefowl, although numbers have not increased in all baited areas. Malleefowl have been successfully translocated to Francois Peron National Park at Shark Bay. Malleefowl conservation is a major undertaking of several non-government organisations. The success of the Malleefowl Preservation Group, based at Ongerup, has led to the formation of similar groups at Wubin-Dalwallinu, Morawa, Kalgoorlie and elsewhere.

Below Malleefowl have declined because of clearing and fox predation of young birds.

Photo – Ken Stepnell/CALM

Recherche Cape Barren goose
Cereopsis novaehollandiae grisea

Other names: Cape Barren goose (WA subspecies).

Description: This large (75–100 cm), grey, goose-like bird has dark spots on its wing coverts, black tips on its wings and a dark tail. Its short, black bill has a prominent lime green cere (soft swelling at the base of the upper beak that contains the nostrils). Legs are red and the feet are black. Another, more common, subspecies occurs on islands off South Australia and in Bass Strait.

Distribution and habitat: The Recherche Cape Barren goose breeds only on islands in the Archipelago of the Recherche and on a few nearby islands. It resides on islands but is occasionally seen on the mainland near Esperance, especially in summer. The population is very small. A helicopter survey by the Department of Conservation and Land Management in April 1993 sighted 631 geese, a figure very near the total number.

Biology and ecology: Cape Barren geese are herbivorous, eating grasses and herbs. They build a saucer-shaped nest, lined with grass and down, on the ground or on a low bush. They lay four to six eggs from June to August. The young are flying by October.

Threats: This subspecies is naturally rare. Potential threats include the introduction of predators to its breeding islands. Climate change may have a profound effect, as many birds died during a very hot spell in early 1991.

Status: Vulnerable. All breeding islands are nature reserves.

Above The naturally rare western subspecies of the Cape Barren goose breeds in the Archipelago of the Recherche, near Esperance.

Photo – Jiri Lochman

Above Southern giant-petrel on Macquarie Island.

Photo – O Ertok,
Australian Antarctic Division,
Commonwealth of Australia

Southern giant-petrel
Macronectes giganteus

Description: This very large, albatross-sized petrel varies in colour from very dark to almost white, and has a bulbous beak. Giant-petrels appear to be very clumsy when sitting on water or on land, but fly and glide effortlessly when on the wing. This species is very difficult to distinguish from the northern giant-petrel, which has a reddish-brown rather than a green-tipped beak and a darker head and neck in adult birds.

Distribution and habitat: Southern giant-petrels breed on Macquarie and Heard islands, and on other subantarctic islands. They are widespread in southern oceans, and have been recorded as far north as Shark Bay in WA.

Biology and ecology: They nest annually in small colonies amongst open vegetation, but about 30% of the adult population does not breed in any one year. A single chick is raised. Southern giant-petrels feed on squid and krill, and will follow fishing boats looking for scraps.

Threats: Longline fishing is the major threat (see threatened albatrosses on p. 87).

Status: Vulnerable. Albatrosses and giant-petrels are the subject of a national recovery plan.

White-chinned petrel   *Procellaria aequinoctialis*

Below White-chinned petrels breed on subantarctic islands.

Photo – Hans and Judy Beste
/Lochman Transparencies

Other names: Cape hen, spectacled petrel, shoemaker.

Description: White-chinned petrels are dark in colour, except for a variable white chin patch. The birds are 51–58 cm long, with a wingspan of 134–137 cm. They are distinguished from other petrels and shearwaters by the all-pale bill and black feet.

Distribution and habitat: The species breeds on subantarctic islands and wanders widely in southern oceans.

Biology and ecology: White-chinned petrels feed on fish, crustaceans and squid, and particularly on offal. They nest colonially in burrows on subantarctic islands. The largest population is on South Georgia Island.

Threats: White-chinned petrels are threatened by longline fishing (see threatened albatrosses on p. 87).

Status: Vulnerable.

Threatened albatrosses

Thirteen species of albatrosses are listed as threatened in WA. These are the Amsterdam albatross (*Diomedea amsterdamensis*), Tristan albatross (*Diomedea dabbenena*), southern royal albatross (*Diomedea epomophora*), wandering albatross (*Diomedea exulans*), Gibson's albatross (*Diomedea gibsoni*), northern royal albatross (*Diomedea sanfordi*), sooty albatross (*Phoebetria fusca*), light-mantled albatross (*Phoebetria palpebrata*), Indian yellow-nosed albatross (*Thalassarche carteri*), shy albatross (*Thalassarche cauta*), Atlantic yellow-nosed albatross (*Thalassarche chlororhynchos*), grey-headed albatross (*Thalassarche chrysostoma*) and Salvin's albatross (*Thalassarche salvini*).

Distribution and habitat: Albatrosses breed on subantarctic and other southern ocean islands and fly enormous distances in the southern oceans searching for food.

Biology and ecology: Albatrosses, being very large birds, breed late in life and usually raise only a single young every two years.

Threats: All southern hemisphere albatross species, as well as giant-petrels and some other seabirds, are threatened by the longline fishing industry. All species take baited hooks, usually when they are being set and before they have sunk, and then drown. Fishing technology has been developed to prevent albatrosses and other seabirds from ingesting baited hooks, but—while these are mandatory on Australian boats and boats fishing in Australian and some other nations'
waters—they are not widely used by other countries' fishing fleets. Some island breeding colonies are also threatened by introduced predators and other introduced animals.

Above Light-mantled albatross and chick on Macquarie Island.

Photo – O Ertok, Australian Antarctic Division, Commonwealth of Australia

Above Royal albatross on Middle Sisters Island, New Zealand.
Photo – G Johnstone, Australian Antarctic Division, Commonwealth of Australia

Above right Salvin's albatross.
Photo – Hans and Judy Beste /Lochman Transparencies

Below Grey-headed albatross and chick on Macquarie Island.
Photo – G Johnstone, Australian Antarctic Division, Commonwealth of Australia

Below right Wandering albatross on Macquarie Island.
Photo – C Baars, Australian Antarctic Division, Commonwealth of Australia

The statuses shown here are the worldwide status of each species. The status of some albatross populations breeding within Australian territory is worse. For example, there were 44 breeding pairs of wandering albatrosses on Macquarie Island in 1967/68, but according to the latest report less than ten pairs remain. Australia is at the forefront of international efforts to reduce the threat of longline fishing to albatrosses and other seabirds. Conservation efforts are made difficult because most longline fishing takes place in international waters, by boats registered in countries that will not cooperate with conservation efforts.

Status:

Amsterdam albatross	Critically Endangered
Tristan albatross	Endangered
southern royal albatross	Vulnerable
wandering albatross	Vulnerable
Gibson's albatross	Vulnerable
northern royal albatross	Endangered
sooty albatross	Vulnerable
light-mantled albatross	Vulnerable
Indian yellow-nosed albatross	Vulnerable
shy albatross	Vulnerable
Atlantic yellow-nosed albatross	Vulnerable
grey-headed albatross	Vulnerable
Salvin's albatross	Vulnerable

Above Shy albatross. Photo – Brett Dennis/Lochman Transparencies

Above Masked boobies.

Photo – Andrew Burbidge /CALM

Masked booby (eastern Indian Ocean)   *Sula dactylatra bedouti*

Other names: Masked gannet, blue-faced gannet/booby.

Description: This striking white booby, 80–85 cm long, has a black face, black flight feathers and a black tail. The bill is yellow.

Distribution and habitat: In WA, the masked booby nests on Adele and Bedout islands and West Island in the Lacepede group. It also nests on North Keeling (about 30 pairs) and West and Middle islands on Ashmore Reef, and in Indonesia. Counts suggest populations of 300–400 breeding pairs on Adele, 120–270 pairs on Bedout and less than 10 pairs on West Lacepede Island. Very few birds breed on the Ashmore Reef islands, probably due to illegal hunting by Indonesian fishers.

Biology and ecology: These birds nest in small colonies. The nest, in which two eggs are laid, is a scrape on the ground. Masked boobies feed on fish.

Threats: Masked boobies are thought to be naturally rare. Threats include illegal killing on nesting islands, and over-harvesting of fish stocks on which they depend for food.

Status: Vulnerable. All nesting islands in WA are nature reserves. All breeding islands require protection from illegal hunting by Indonesian fishers. The Commonwealth Government monitors illegal fishing in Australian waters and many boats and crews have been arrested and prosecuted.

Cape gannet   *Morus capensis*

Other names: South African gannet.

Description: The Cape gannet is very similar to the Australasian gannet, but has a completely black tail and a longer throat stripe. It is 84–94 cm long, with a 171–185 cm long wingspan.

Distribution and habitat: Most birds breed off South Africa, but Cape gannets recently started to breed in Australia in Port Phillip Bay, Victoria, and are occasionally reported off WA. The South African population is in rapid decline, with a 31% reduction between 1956 and 2000.

Biology and ecology: Gannets plummet into the sea from a great height and plunge to a considerable depth to catch fish. A single egg is laid on a mound of seaweed and guano.

Threats: The principal threat to the small Australian population is hybridisation with the Australasian gannet (*Morus serrator*). Whether this requires management action is questionable.

Status: Vulnerable.

Australalasian bittern
Botaurus poiciloptilus

Other names: Brown bittern, bunyip bird (English), burdenedj/bardanidj (Nyoongar).

Description: This is the largest Australian bittern, reaching 70 cm long. Its plumage is mostly in various shades of brown, allowing it to camouflage itself among reeds and rushes. It has a loud, deep, resonant, bray-like call.

Distribution and habitat: The Australalasian bittern inhabits shallow freshwater swamps with fairly dense low vegetation of reeds, sedges, rushes and grasses. In WA it is confined to the south-west corner, but it also occurs in south-eastern Australia (including Tasmania), New Zealand and New Caledonia. It has disappeared from the Wheatbelt, and is now largely confined to coastal areas, especially along the south coast. The Lake Muir-Lake Unicup wetlands, which are mostly within nature reserves, are a stronghold.

Biology and ecology: The species hunts mainly at night and feeds primarily on large aquatic invertebrates and small vertebrates such as tadpoles, frogs and fish. The nest is a platform of reeds and rushes built in dense vegetation. Breeding takes place in late winter and spring.

Threats: Loss of habitat due to clearing, drainage and salinisation are the main threats to the Australalasian bittern. Altered fire regimes may have also contributed to its decline.

Status: Vulnerable.

Above Now absent from the Wheatbelt, the Australalasian bittern is confined to the south-west corner of Western Australia.

Photo – Hans and Judy Beste /Lochman Transparencies

Below Red goshawk.

Photo – D Hollands /NatureFocus

Red goshawk   *Erythrotriorchis radiatus*

Other names: Red or rufous-bellied buzzard.

Description: This robust, medium-sized hawk is 46–61 cm long. Its crown, nape, breast and wings are rust coloured, streaked with black. The face and throat are paler. The long wings are tipped with black.

Distribution and habitat: In WA, the species is found mainly in medium-density woodlands and forests of the Kimberley, but it also occurs across northern Australia and near the east coast.

Biology and ecology: The red goshawk is the largest and most powerful bird-eating hawk in Australia. It takes large birds such as cockatoos and ducks as well as nestlings, small mammals, snakes and other reptiles. The nest is built from large sticks well off the ground.

Threats: The red goshawk is naturally rare, and has declined near the east coast. It is very rare in the Kimberley, but there is no evidence of numbers having declined there. Numbers probably total less than 1000. There are no obvious threats in WA.

Status: Vulnerable.

Houtman Abrolhos painted button-quail   *Turnix varia scintillans*

Other names: Speckled or varied quail, painted button-quail (Houtman Abrolhos).

Description: The Houtman Abrolhos painted button-quail has a rich rufous, black and grey pattern on its back and sides, and a grey breast with silver marbling. It is 14–17 cm long. Females are larger and more brightly coloured than males.

Distribution and habitat: This subspecies occurs only in the Wallabi group of the Houtman Abrolhos Islands, where it has been recorded on North Island and East and West Wallabi islands, and on some small islands adjacent to the Wallabi islands. It occupies most habitats—low coastal scrub, beach spinifex (*Spinifex longifolius*), grassland and saltmarsh—on the islands.

Biology and ecology: Button-quail are polyandrous—the female takes up a territory to which she attracts a mate, who incubates the eggs and raises the young while she seeks up to three more territories and mates. Button-quail feed on seeds and insects.

Threats: The introduction of predators or competitors and frequent or extensive fires are potential threats. House mice and tammar wallabies have been introduced to North Island. Over-grazing by the wallabies may be affecting the quail's food plants.

Status: Vulnerable. All islands in the Houtman Abrolhos are reserved, and there are proposals that most be made into a national park.

Australian painted snipe   *Rostratula australis*

Other names: Painted snipe.

Description: The Australian painted snipe is 22–25 cm long. It has a mottled green, grey and buff back, stripes on its head and a long, drooping pink bill. The legs are dull green. The female has a chestnut chin and neck.

Distribution and habitat: This species is found only in Australia. It has been widely but rarely recorded in WA, where it inhabits shallow, vegetated swamps.

Biology and ecology: The female mates with more than one male during the breeding season. She lays three to six eggs, which are incubated by the male in a shallow scrape. The species is reported to feed on vegetation, seeds and invertebrates.

Threats: Drainage and diversion of water from temporary swamps have reduced available habitat, and overgrazing of shallow swamps may also have affected the species. Recent studies suggest that it has become very rare and may be Critically Endangered.

Status: Vulnerable.

Lesser noddy (Australian subspecies)
Anous tenuirostris melanops

Other names: Lesser noddy (eastern Indian Ocean).

Description: This dark-coloured tern, 29.5–34.5 cm long, is very similar to the common (or brown) noddy (*Anous stolidus*), but is slightly smaller, greyer and has light grey, rather than black, feathers between the eye and the base of the beak.

Distribution and habitat: The Australian subspecies breeds only on three islands in the Houtman Abrolhos: Pelsaert Island in the Southern group and Wooded and Morley in the Easter group. Monitoring of colony size by Department of Conservation and Land Management staff, in association with the Department of Fisheries, since 1989 has revealed there are between 35,000 and 40,000 breeding pairs on Pelsaert Island, 5000 and 15,000 on Wooded and 8000 and 16,000 on Morley. Despite being a relatively abundant bird, the subspecies is dependent on less than 4.5 ha of mangroves for breeding, and hence is very susceptible to habitat destruction. Another subspecies breeds on islands in the western Indian Ocean.

Biology and ecology: Nesting is colonial. Nests are constructed from seaweed and placed on a horizontal mangrove branch. Lesser noddies feed on small fish and squid during the day, roosting in mangroves at night. They remain at the Abrolhos all year round.

Threats: Introduction of predators (especially rats), death of mangroves due to pollution (from an oil spill, for example) or flooding should the sea level rise significantly due to global warming, and depletion of food through over-fishing are all potential threats. The islands on which it breeds are reserved and managed by the Department of Fisheries. All lesser noddy breeding islands are within a proposed national park.

Status: Vulnerable.

Above The Australian subspecies of the lesser noddy breeds only in 4.5 ha of mangroves on three islands in the Houtman Abrolhos, offshore from Geraldton.

Photo – Andrew Burbidge /CALM

Below Subantarctic skua.

Photo – Eric Woehler, Polar-eyes.com

Subantarctic skua (southern)   *Catharacta lonnbergi lonnbergi*

Other names: *Catharacta skua lonnbergi*, great skua, southern skua.

Description: This large (60–64 cm long), stocky, gull-like bird is dark brown, with white bases to the primary feathers. It has a black bill and a short tail.

Distribution and habitat: The subantarctic skua breeds on Macquarie and Heard islands and possibly on Macdonald Island, all in Australian territory, as well as on other subantarctic islands. There is probably little or no genetic interchange between populations.

Biology and ecology: Skuas are predators and scavengers, feeding among penguin and seal colonies and also capturing live food. Rabbits are eaten on Macquarie Island.

Threats: The subantarctic skua is naturally rare. Its numbers are affected by numbers of penguins and seals, so over-fishing of these species' food would also affect the skua.

Status: Vulnerable.

Above The western subspecies of the partridge pigeon can be observed in the Mitchell River National Park.

Below The long upper mandible is used by Baudin's black-cockatoo to extract seeds from marri nuts.

Photos – Jiri Lochman

Partridge pigeon (western) *Geophaps smithii blaauwi*

Other names: Western partridge pigeon.

Description: The partridge pigeon is 21.5–28 cm in length. The upper parts and breast are brown. The belly and flanks are white, and there is a broad patch of yellow skin around the eye. The eastern subspecies (*Geophaps smithii smithii*) has a red eye patch and once occurred near Kununurra, but is now restricted to the Northern Territory.

Distribution and habitat: The western subspecies of the partridge pigeon is restricted to coastal areas between Kalumburu and Kimbolton in the Kimberley. It inhabits open forest and savannah woodland with an open understorey.

Biology and ecology: The birds forage on the ground, eating seeds of grasses, legumes and herbs. Two white eggs are laid on the ground, in a shallow depression lined with leaves or grass.

Threats: Even though it lives in remote areas, changed fire regimes, predation by feral cats and habitat degradation by feral cattle and donkeys all threaten this rare subspecies.

Status: Vulnerable.

Baudin's black-cockatoo *Calyptorhynchus baudinii*

Other names: Long-billed black-cockatoo.

Description: This large black-cockatoo has white under its tail. It is distinguished from Carnaby's black-cockatoo by the long, downcurved tip on its upper mandible, which is used to extract seeds from fruits, especially from marri (*Corymbia calophylla*).

Distribution and habitat: Baudin's black-cockatoo is restricted to the south-west corner of WA from near Perth to Albany, inland to about Narrogin. It inhabits jarrah (*Eucalyptus marginata*), marri and karri (*E. diversicolor*) forest, woodland and coastal scrub.

Biology and ecology: This bird nests in large, deep hollows in karri, marri and wandoo (*E. wandoo*), where two eggs are laid. Breeding success, however, is low. Nests are thinly dispersed. Baudin's black-cockatoo feeds mainly on marri seeds and flowers, but will take seeds from apples and pears in orchards and from pine cones.

Threats: Up to a quarter of the original habitat of Baudin's black-cockatoo has been cleared for agriculture. Logging has removed many of the old trees with good hollows (although it is not known to what extent this is significant because of the thinly dispersed nests) and has also removed many big marri that produce large quantities of flowers and seeds. The reduction in marri logging since 2001 should reduce this threat.

Status: Vulnerable. Although there is little evidence of a continuing decline, the Western Australian Threatened Species Scientific Committee recommended this status under the precautionary principle.

Carnaby's black-cockatoo *Calyptorhynchus latirostris*

Other names: Short-billed black-cockatoo (English), manatj, ngoorlak (ngoolyanak refers to a flock of black-cockatoos) (Nyoongar). Some or all Nyoongar names may also apply to Baudin's black-cockatoo.

Description: Carnaby's black-cockatoo is a large, black cockatoo with white under its tail. It is distinguished from Baudin's black-cockatoo by its shorter upper mandible.

Distribution and habitat: Carnaby's black-cockatoos are confined to the south-west of WA. They breed in the northern kwongan (shrubby heathlands) and Wheatbelt, but spend summer near the west and south coasts. They are rare in, or absent from, forested areas.

Biology and ecology: The birds nest in large, deep hollows, mainly in salmon gum (*Eucalyptus salmonophloia*) and wandoo (*E. wandoo*). Two eggs are laid, but only one chick usually survives. They feed in kwongan. Seeds of hakeas, banksias, grevilleas and eucalypts form the bulk of their diet, but insects, particularly boring insects, are also eaten. They also feed in pine plantations. Carnaby's black-cockatoos have been studied over many years by Denis Saunders and his colleagues from CSIRO.

Threats: Clearing, and subsequent land degradation, has eliminated most of the breeding habitat and much of the coastal kwongan where the Carnaby's black-cockatoo spends its time in summer. As it requires old trees with large hollows in which to nest, it will take many decades for trees planted now to become suitable. Because of clearing, the distances between remaining nesting trees and feeding areas in kwongan have become so great in some areas that the chicks starve. Competition for nesting hollows by increasing numbers of galahs

Above Most of the breeding habitat of Carnaby's black-cockatoo has been destroyed by clearing.

Photo – Bill Belson /Lochman Transparencies

(*Cacatua roseicapilla*), western corellas (*C. pastinator butleri*) and feral honey bees may be significant. Illegal trapping, which often involves the destruction of nesting hollows, has been a threat and, although demand from the domestic market has been largely eliminated via effective enforcement, smuggling is still an issue. Extensive harvesting of mature pines may reduce the summer food supply.

Status: Endangered. Implementation of a recovery plan, overseen by a recovery team, is underway.

Above Muir's corellas are threatened by loss of nesting trees.

Photo – Jiri Lochman

Below Western ground parrots.

Illustration – Martin Thompson

Muir's corella   *Cacatua pastinator pastinator*

Other names: Western corella, western long-billed corella (English), manait, moonit, manap (Nyoongar).

Description: Muir's corella is mostly white, with pinkish-red between the eye and bill and on the throat. It has a long, slender upper bill and black feet. The bird is 38–41 cm long.

Distribution and habitat: Once reasonably common between Perth and Albany, Muir's corella is now limited to a small area near Boyup Brook and around Lake Muir. It inhabits open forests and woodlands.

Biology and ecology: The Muir's corella nests in large hollows in old eucalypts, and uses its long upper bill to dig for roots, tubers and corms. Because most of its range is now cleared, it feeds mostly on introduced plants such as Guildford grass (*Romulea* spp.).

Threats: The main threat is the continued loss of old nesting trees, which are being removed to plant bluegums, vegetables and vines. The trees are also dying of old age. Once extensively poisoned and shot, Muir's corellas are still illegally killed by some landholders, as they feed on young cereal crops and grain. In the longer term, they are threatened by hybridisation with introduced eastern long-billed corellas (*C. tenuirostris*).

Status: Endangered. A recovery team has prepared a recovery plan. Elimination of eastern long-billed corellas, established in metropolitan Perth from cage bird escapes, is important.

Western ground parrot   *Pezoporus wallicus flaviventris*

Other names: Ground parrot (western subspecies) (English), boorandadi, djadong, djadoonkari, djoolbada, kailoring (Nyoongar).

Description: This smallish green parrot is around 30 cm long. Its upper parts are mottled black and yellow, its underparts are yellow with short, blackish bars and it has a long greenish tail. Adults have a distinct red patch at the base of the bill.

Distribution and habitat: Western ground parrots once occurred to the north of Perth, as well as near Albany. They are now confined to the south coast at Waychinicup, Fitzgerald River and Cape Arid national parks. Ground parrots live in low species-rich coastal kwongan (shrubby heath), no more than 50 cm high.

Biology and ecology: Although they can fly well, western ground parrots spend most of their time on the ground, where they feed on seeds. Nesting also takes place on the ground in a shallow hollow under a bush.

Threats: Clearing and stock grazing have eliminated a lot of habitat in the past. Frequent burning is a serious threat, as birds are most plentiful in long-unburnt kwongan and do not reinvade regenerating kwongan for several years. Extensive fire is also a threat, especially in Fitzgerald River and Cape Arid national parks. Because the parrots live on the ground, predation by cats and foxes is a serious threat.

Status: Endangered. Western ground parrots are close to qualifying as Critically Endangered. Their recovery is being managed by the WA South Coast Threatened Birds Recovery Team.

Night parrot *Pezoporus occidentalis*

Other names: Spinifex parrot.

Description: Night parrots are 23 cm long. This species is similar to the ground parrot, but lacks red on the face.

Distribution and habitat: This species has been recorded from pastoral and desert areas of WA, the Northern Territory, South Australia, Queensland and New South Wales. There were many records from inland Australia in the 1870s and 1880s, and the type specimen came from the Murchison. Several reliable reports came from the Murchison and Gascoyne early in the twentieth century, but there have been no confirmed sightings of night parrots since the 1930s. The species was considered by many to be extinct until a recently-dead bird was found in western Queensland in 1990.

Biology and ecology: Night parrots inhabit spinifex (*Triodia*) hummock grasslands or chenopod shrublands, especially around salt lakes. They probably feed mainly on seeds and green herbage. They nest at the end of a tunnel in dense vegetation, such as a spinifex hummock.

Threats: Predation—especially of nests— by cats and foxes, changed fire regimes in spinifex and grazing of shrublands are thought to have brought the night parrot to the brink of extinction.

Status: Critically Endangered. There have been many searches for this enigmatic species. An active campaign by the Department of Conservation and Land Management, targeting mainly outback residents and workers, to try to locate a population in WA, has so far been unsuccessful.

Below The Critically Endangered night parrot has not been found alive since the 1930s.

Illustration – William T Cooper, by permission of the National Library of Australia.

Above The recovery of the noisy scrub-bird, once thought to be extinct, is a conservation success story.

Opposite Noisy scrub-bird.

Photos – Babs and Bert Wells /CALM

Noisy scrub-bird *Atrichornis clamosus*

Other names: The Nyoongar name is recorded as jee-mul-uk. Depending on pronunciation of the 'u', spelling according to the rules of the Nyoongar Dictionary would be djimalak or djimoolook.

Description: Noisy scrub-birds are brown, with narrow black lines on the back. They are 20–23 cm in length. The male has a black and white throat and a rufous belly. The species is very secretive and is heard much more often than it is seen.

Distribution and habitat: The species was originally distributed in the Darling Range and along the south-west coast, from near Margaret River to Mount Barker and Albany. There were no sightings between 1889 and 1961, when the noisy scrub-bird was rediscovered at Two Peoples Bay, east of Albany.

Biology and ecology: Noisy scrub-birds live in the transitional vegetation between swamps and forests. On Mount Gardner they live in and adjacent to narrow, wet gullies with dense sedges. A single egg is laid during winter, in a domed nest constructed from sedges. Noisy scrub-birds feed on invertebrates, mostly ants, spiders and beetles, that inhabit the dense leaf litter.

Threats: The Darling Range population was probably wiped out by frequent fires, associated with the then-unregulated timber industry, in the nineteenth century. South coast populations also are thought to have succumbed to the frequent fires lit by

early pastoralists, and to swamp drainage and agricultural clearing. The birds survived on Mount Gardner, at Two Peoples Bay, because it is on a peninsula into which prevailing winds rarely direct fires, and because, when fires do occur, they do not burn out the whole peninsula due to natural firebreaks of bare rock.

Status: Vulnerable. One of the first recovery plans written in Australia was for the noisy scrub-bird, in 1986. It could not have been written without scientific research carried out by Graeme Smith of CSIRO, and training in translocation techniques provided by Don Merton of the New Zealand Department of Conservation. Recovery of the noisy scrub-bird has been a very successful conservation program. When the species was rediscovered in 1961, there were less than 100 birds remaining in one small area. Now there are five coalescing populations in the Two Peoples Bay and Mount Manypeaks area, near Albany, with about 800 breeding pairs. Populations are also being reestablished in the Darling Range near Harvey. The primary strategy has been to establish several populations, so that fire can affect only a relatively small number at the one time. This strategy proved successful when, in late 2000, a major fire at Two Peoples Bay destroyed about 40 scrub-bird territories. This was such a small proportion of the current total that it was not necessary to reevaluate the conservation status of the species.

Shark Bay variegated fairy-wren *Malurus lamberti bernieri*

Description: The Shark Bay variegated fairy-wren differs from mainland variegated fairy-wrens by having a much darker bluish-violet cap and cobalt blue ear tufts.

Distribution and habitat: The subspecies is restricted to Bernier and Dorre islands in Shark Bay. Dorre Island birds appear to intergrade with the mainland subspecies (*M. lamberti assimilis*).

Biology and ecology: Variegated fairy-wrens live in low heath and among shrubs. They eat insects and occasionally spiders, seeds and leaves.

Threats: Introduction of predators and extensive or frequent fire are potential threats.

Status: Vulnerable. Bernier and Dorre islands are both nature reserves.

Dirk Hartog Island black-and-white fairy-wren
Malurus leucopterus leucopterus

Other names: White-winged fairy-wren (Dirk Hartog Island).

Description: The birds are 11.5–13 cm long. This species is similar to the white-winged fairy-wren of mainland Australia, but is black and white rather than blue and white. Recent genetic studies have suggested that Dirk Hartog Island black-and-white fairy-wrens are so closely related to mainland white-winged fairy-wrens that they may not warrant subspecific status.

Distribution and habitat: The subspecies is widespread on Dirk Hartog Island, in Shark Bay, and occupies most habitats.

Biology and ecology: Fairy-wrens are insectivorous and live in small related groups.

Threats: Dirk Hartog Island is probably too large for one or even a series of fires to greatly harm the fairy-wrens, however, the possible introduction of rats is a major concern. The fairy-wrens, along with several other species of small passerine birds, have survived more than 100 years of pastoralism and the introduction of cats and mice. Thick-billed grasswrens (*Amytornis textilis textilis*), however, have become extinct on the island.

Status: Vulnerable. Dirk Hartog Island, currently a pastoral lease, will soon become a national park, hence improving the status of Dirk Hartog Island black-and-white fairy-wrens.

Below This striking black and white subspecies of the white-winged fairy-wren is restricted to Dirk Hartog Island, Shark Bay.

Photo – Babs and Bert Wells /CALM

Barrow Island black-and-white fairy-wren
Malurus leucopterus edouardi

Other names: White-winged fairy-wren (Barrow Island).

Description: In appearance, this subspecies is almost indistinguishable from the Dirk Hartog Island birds. Recent genetic studies have shown that the Barrow Island black-and-white fairy-wren is more distantly related to the black-and-white birds on Dirk Hartog Island than to the mainland white-winged fairy-wren.

Distribution and habitat: They are restricted to Barrow Island in the Pilbara. Black-and-white fairy-wrens, presumably of this subspecies, may also have occurred in the Montebello Islands. As the possible sighting was on Trimouille Island in 1950, the nuclear weapon test there in 1952 may have wiped it out.

Biology and ecology: Barrow Island fairy-wrens inhabit spinifex (*Triodia*) hummock grassland and are insectivorous.

Dirk Hartog Island southern emu-wren
Stipiturus malachurus hartogi

Description: Dirk Hartog Island southern emu-wrens are similar to mainland southern emu-wrens, but paler.

Distribution and habitat: This subspecies is restricted to Dirk Hartog Island in Shark Bay.

Biology and ecology: Emu-wrens inhabit low, dense vegetation, and are insectivorous.

Threats: See Dirk Hartog Island black-and-white fairy-wren on p. 100.

Status: Vulnerable.

Threats: Barrow Island fairy-wrens have survived very large fires in the past, as the island is too large for a single fire to wipe them out. The introduction of rats, cats or other predators is the greatest threat.

Status: Vulnerable. Barrow Island is a nature reserve. An oilfield operated by ChevronTexaco covers the southern half of the island.

Above Barrow Island black-and-white fairy-wren.

Photo – Steve Pruett-Jones

Below A male southern emu-wren from Dirk Hartog Island.

Photo – Babs and Bert Wells /CALM

Above Good fire management is needed to ensure the conservation of the western bristlebird.

Photo – Simon Nevill /Lochman Transparencies

Western bristlebird   *Dasyornis longirostris*

Other names: Long-billed or brown bristlebird.

Description: This medium-sized (17 cm long), ground-dwelling bird is sturdy, shy and elusive. It is mainly dark brown above (the top of its head and neck are dark brown, with flecks of silvery grey). The wings, rump and tail are rufous brown, whereas the underparts are off-white to brownish. The noisy scrub-bird has similar colouration, but its upper parts are barred with black, it lacks grey flecking and has a very different call.

Distribution and habitat: Once found in coastal areas from Perth to Augusta, and from near Albany to the eastern end of Fitzgerald River National Park, western bristlebirds are now found only in and between Two Peoples Bay Nature Reserve and Waychinicup National Park, with remnant populations in the Fitzgerald River National Park.

Biology and ecology: Western bristlebirds prefer low, dense kwongan (shrubby heath), sometimes with scattered patches of mallee or other stunted trees.

Threats: Land clearing has eliminated much of their habitat. Extensive fire is the current major threat in remaining habitat. Western bristlebirds can escape fires so long as areas of habitat remain. They do not reinvade a burnt area for several years, with the number of years depending on rainfall. Swampy areas are recolonised within two to three years, kwongan at Two Peoples Bay is used again within three to five years, while kwongan at Fitzgerald River National Park is not used for 14 to 28 years. Management burning (apart from that associated with control or management of wildfire) is not required, as birds at Two Peoples Bay still live in areas unburnt for more than 50 years.

Status: Vulnerable. However, the species is close to qualifying as Endangered, and would require reassessment should significant areas of habitat be burnt. The South Coast Threatened Birds Recovery Team coordinates conservation actions. Considerable research has been conducted at Two Peoples Bay by CSIRO and the Department of Conservation and Land Management, with continuing research at all known locations. A translocation to the Nuyts Wilderness area of Walpole-Nornalup National Park in 1999 and 2000 was successful, but most western bristlebird habitat there was burnt in a very hot fire in 2001. At least seven birds escaped the fire and now live in adjacent habitat.

Dirk Hartog Island rufous fieldwren   *Calamanthus campestris hartogi*

Other names: Dirk Hartog Island calamanthus.

Description: Rufous fieldwrens are about 12 cm long and are streaked brown above. They are paler below, with marked streaking on the breast. There is a light eyebrow stripe. The birds typically sit with the tail cocked upwards. This subspecies is paler above and lighter below than the adjacent mainland subspecies (*C. campestris rubiginosus*).

Distribution and habitat: The subspecies is restricted to Dirk Hartog Island in Shark Bay. Fieldwrens live in low, sparse heath, saltmarsh and samphire.

Biology and ecology: Fieldwrens eat insects and lay two or three eggs in a domed nest built near the ground.

Threats: See Dirk Hartog Island black-and-white fairy-wren on p. 100.

Status: Vulnerable.

Dorre Island rufous fieldwren   *Calamanthus campestris dorrie*

Other names: Dorre Island calamanthus.

Description: This subspecies is very similar to subspecies *hartogi* (see above).

Distribution and habitat: The Dorre Island rufous fieldwren is restricted to Dorre Island in Shark Bay.

Biology and ecology: On Dorre Island, rufous fieldwrens live in low, sparse vegetation and feed on insects and seeds.

Threats: The introduction of predators, especially rats or cats, and extensive or frequent fire are potential threats.

Status: Vulnerable. Dorre Island is a nature reserve.

Above The western whipbird is a threatened species surviving at Two Peoples Bay Nature Reserve near Albany.

Photo – Babs and Bert Wells /CALM

Western whipbird (western heath subspecies)
Psophodes nigrogularis nigrogularis

Description: This medium-sized (23.5 cm long), ground-dwelling bird has a short, erect, triangular crest on its forehead and long, powerful legs. It is mostly olive above with a prominent white 'moustache' stripe extending from behind the beak and along the neck. It has a black chin, a grey breast and an off-white belly. It differs from subspecies *oberon*, which occurs in mallee heaths in the Stirling Range and to the east of Cheyne Beach, by its shorter wings and tail, its off-white rather than white belly and its narrower white 'moustache' stripe.

Some authorities consider that the western heath subspecies of the western whipbird should be treated as a full species, but DNA analyses carried out by Les Christidis and Janette Norman of the Museum of Victoria, funded by the Department of Conservation and Land Management, do not support this contention.

Distribution and habitat: The western heath subspecies of the western whipbird once occurred near the west coast, from Perth to Augusta, and the south coast, from Albany east to at least Two Peoples Bay. It has disappeared from the west coast and is now restricted to a small area east of Albany, between Mount Taylor and Cheyne Beach. Most birds are in Two Peoples Bay Nature Reserve and Waychinicup National Park.

Biology and ecology: The western whipbird inhabits dense shrubland with an open understorey. It builds a domed nest in dense bushes in heath adjacent to thickets. The birds feed on the ground and are insectivorous. The Two Peoples Bay population has increased from about 60 pairs in 1970 to about 100 pairs in 1982.

Threats: Clearing for agriculture and changes in fire regimes probably caused the original decline. Frequent and extensive fire is the major current threat and is thought to have caused the disappearance of the species from the Leeuwin-Naturaliste ridge.

Status: Vulnerable. Conservation is coordinated by the South Coast Threatened Birds Recovery Team. Fire management of Two Peoples Bay Nature Reserve and Waychinicup National Park is important for this species, as well as for several other threatened birds and Gilbert's potoroo.

Northern crested shrike-tit
Falcunculus frontatus whitei

Other names: Northern shrike-tit (English), bunbun (Wunambal).

Description: This subspecies is around 17 cm long. It is thick set, with a large head and a heavy wedge-shaped bill. It has a short erect crest, bold black and white stripes on its head and neck, olive upper parts, and mainly yellow underparts. It is not easily confused with any other bird within its range.

Distribution and habitat: The northern subspecies of the crested shrike-tit, as its names implies, occurs in northern Australia, including the Kimberley. It inhabits woodlands, especially those dominated by Darwin woollybutt (*Eucalyptus miniata*), Darwin stringybark (*E. tetrodonta*) and smooth-stemmed bloodwood (*Corymbia bleeseri*).

Biology and ecology: Little is known about this extremely rare bird. There were only 26 records from 22 localities across the Northern Territory and Kimberley up to 1992, and there have been few since. It is assumed that—like the two other subspecies of crested shrike-tit—they glean insects and other invertebrates from beneath ribbons of bark that peel from certain eucalypts. They build cup-shaped nests in a tree fork and lay two to three eggs.

Threats: Increased frequency and intensity of fires in savannah woodlands in northern Australia, especially late in the dry season, are thought to have reduced the availability of food.

Status: Endangered. Research is needed to locate populations and develop conservation actions.

Right A crested shrike-tit of the south-western subspecies.

Photo – Babs and Bert Wells/CALM

Above The striking Gouldian finch is common in captivity but threatened with extinction in the wild.

Photo – Ken Stepnell/CALM

Gouldian finch   *Erythrura gouldiae*

Description: The Gouldian finch is around 14 cm long. It is green above, with a blue rump. The male has a lilac chest and yellow abdomen, an ivory-coloured bill with a red tip and a black face (a few birds have crimson or yellowish-orange faces). Females are duller.

Distribution and habitat: Gouldian finches occur from the Kimberley across northern Australia to Cape York. Their current range is similar to their past range in WA, but has shrunk in Queensland. However, numbers have declined enormously, especially since the 1970s. Gouldian finches inhabit open savannah woodlands, often in hilly areas.

Biology and ecology: These birds nest in tree hollows. They breed in the dry season and lay large clutches of more than five eggs. Though they may nest several times in a season, few chicks are raised to independence. The finches feed on grass seeds and, as they need to drink regularly in hot weather, live near water.

Threats: Many Gouldian finches are infected with a parasitic mite that lodges in their lungs and this was thought to be the primary cause of their decline. However, it is now considered that increased mortality due to the mites is indicative of stress caused by major habitat changes that have also affected other granivorous species. Cattle grazing and changed fire regimes are thought to be the principal causes of these changes, which reduce food availability.

Status: Endangered. Prevention of large, late, dry-season fires would benefit Gouldian finches and many other savannah-inhabiting animals. Strategic prescribed burning, including aircraft ignition, early in the dry season, appears to be the only way to prevent large late-season fires in much of northern Australia.

References

Blyth, J. (1996). Night Parrot (*Pezoporus occidentalis*) Interim Recovery Plan for Western Australia. In Pryde, J., Brown, A. and Burbidge, A. (eds), Interim recovery plans 4–16 for Western Australian critically endangered plants and animals. Wildlife Management Program No. 29. Department of Conservation and Land Management, Wanneroo.

Blyth, John and Burbidge, Allan. (1999). *Threatened and rare birds of Western Australia*. Department of Conservation and Land Management, Como.

Burbidge, A.H. (1996). Western Ground Parrot (*Pezoporus wallicus flaviventris*) Interim Recovery Plan. In Pryde, J., Brown, A. and Burbidge, A. (eds), Interim recovery plans 4–16 for Western Australian critically endangered plants and animals. Wildlife Management Program No. 29. Department of Conservation and Land Management, Wanneroo.

Burbidge, A.H. (2003). Birds and fire in the Mediterranean climate of south-west Western Australia. Pp. 321–347 in Abbott, I. and Burrows, N. (eds), *Fire in ecosystems of south-west Western Australia*. Backhuys Publishers, Leiden, The Netherlands.

Danks, Alan, Burbidge, Andrew A., Burbidge, Allan H. and Smith, Graeme T. (1996). Noisy Scrub-bird Recovery Plan. Wildlife Management Program No. 12. Department of Conservation and Land Management, Perth.

Garnett, S.T. and Crowley, G.M. (2000). *The action plan for Australian birds 2000*. Environment Australia, Canberra.

Higgins, P.J. and Davies, S.J.J.F. (eds). (1996). *Handbook of Australian, New Zealand and Antarctic birds. Vol. 3. Snipe to pigeons*. Oxford University Press, Melbourne.

Higgins, P.J. (ed). (1999). *Handbook of Australian, New Zealand and Antarctic birds. Vol. 4. Parrots to dollarbirds*. Oxford University Press, Melbourne.

Higgins, P.J., Peter, J.M. and Steele, W.K. (eds). (2001). *Handbook of Australian, New Zealand and Antarctic birds. Vol. 5. Tyrant-flycatchers to chats*. Oxford University Press, Melbourne.

Higgins, P.J. and Peter, J.M. (eds). (2002). *Handbook of Australian, New Zealand and Antarctic birds. Vol. 6. Pardalotes to shrike-thrushes*. Oxford University Press, Melbourne.

Johnstone, R.E. and Storr, G.M. (1998). *Handbook of Western Australian birds. Vol. I. Non-passerines (emu to dollarbird)*. Western Australian Museum, Perth.

Marchant, S. and Higgins, P.J. (coordinators). (1990). *Handbook of Australian, New Zealand and Antarctic birds. Vol. 1. Ratites to ducks*. Oxford University Press, Melbourne.

Marchant, S. and Higgins, P.J. (eds). (1993). *Handbook of Australian, New Zealand and Antarctic birds. Vol. 2. Raptors to lapwings*. Oxford University Press, Melbourne.

Reptiles

Western Australia has a very species-rich reptile fauna, with particularly large numbers of species, especially of lizards, in the warmer arid and semi-arid areas, and the Kimberley. The wetter, cooler south-west corner has a comparatively depauperate reptile fauna, but it includes several endemic species.

Many new reptile species have been discovered and described over the past 40 years. More remain to be identified, particularly with the use of modern molecular techniques such as DNA analysis. A checklist published by the WA Museum in 2001 included 464 species and 74 additional subspecies, and more are being described all the time. The number of species and additional subspecies within each group is shown in the table below.

Australian reptiles have fared much better since European settlement than mammals or birds, with relatively fewer threatened species. Only one reptile, Allan's lerista (*Lerista allanae*) from Queensland, is believed to be extinct, and 47 species and subspecies are threatened.

In WA, 13 species and a further four subspecies are listed as threatened: all six species of marine turtles, a freshwater tortoise, nine lizards and a python. One species is listed as Critically Endangered, two as Endangered and 14 as Vulnerable. The proportion of species listed as threatened is 2.8%, and the proportion of species and subspecies listed is 3.2%. Fifty-four species and subspecies of reptiles are included in the Department of Conservation and Land Management's Fauna Priority List. Most of these have very small known distributions, but are not listed because there have been insufficient surveys to be reasonably certain that the species do not actually inhabit larger areas.

Above A green turtle on Barrow Island.

Opposite The striking Yinnietharra rock dragon is known only from a very small part of the Ashburton region.

Photos – Andrew Burbidge

Reptile group	Family	No. of WA species	No. of additional subspecies
marine turtles	Cheloniidae and Dermochelyidae	6	
freshwater turtles	Chelidae (Cheluidae)	8	
dragon lizards	Agamidae	45	15
geckoes	Gekkonidae	60	9
legless lizards	Pygopodidae	45	15
skinks	Scincidae	173	24
monitors (goannas)	Varanidae	18	1
blind snakes	Typhlopidae	23	1
pythons	Boidae	9	3
wart snakes	Acrochordidae	1	
colubrid snakes	Colubridae	3	
water snakes	Homalopsidae	1	
elapid snakes	Elapidae	48	6
sea snakes	Hydrophiidae	22	
crocodiles	Crocodylidae	2	
TOTAL		464	74

Worldwide, marine turtles are threatened by over-harvesting of adults, egg harvesting and predation, lights at nesting beaches, entanglement in fishing gear and through turtles taking baited hooks on longlines. WA retains some of the best populations of turtles in the world, however, there are concerns that turtles that breed on our shores are killed when on the high seas or in seas adjacent to islands to our north, where hunting is a major problem. Traditional harvesting by Aboriginal people in the north of the State is also of concern, and steps are now underway to work with Aboriginal people to ensure that the harvest is sustainable. Egg predation by foxes on the mainland is a significant issue that requires further attention.

The western swamp tortoise (*Pseudemydura umbrina*) was once considered to be the world's most endangered turtle or tortoise. Although it is still Critically Endangered, conservation actions have reduced the likelihood of its extinction and other, mostly Asian, freshwater turtles are now considered more threatened. The western swamp tortoise apparently had a very small range at European settlement, mostly within areas that were cleared for agriculture very soon after Perth was settled.

The major processes threatening terrestrial reptiles have been habitat clearing and habitat degradation due to large wildfires. Predation by foxes and cats is of concern for a number of species, such as the western swamp tortoise, but its relative importance for the majority of threatened species is still unknown. Some species are listed because they have very small distributions and could disappear rapidly if a threat appears (see definition of Vulnerable D2 in Chapter 1).

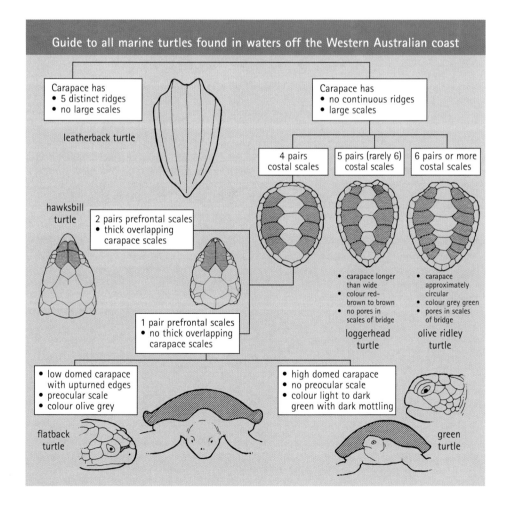

Guide to all marine turtles found in waters off the Western Australian coast

Carapace has
• 5 distinct ridges
• no large scales

leatherback turtle

Carapace has
• no continuous ridges
• large scales

4 pairs costal scales

5 pairs (rarely 6) costal scales

6 pairs or more costal scales

hawksbill turtle

2 pairs prefrontal scales
• thick overlapping carapace scales

1 pair prefrontal scales
• no thick overlapping carapace scales

• carapace longer than wide
• colour red-brown to brown
• no pores in scales of bridge

loggerhead turtle

• carapace approximately circular
• colour grey green
• pores in scales of bridge

olive ridley turtle

• low domed carapace with upturned edges
• preocular scale
• colour olive grey

flatback turtle

• high domed carapace
• no preocular scale
• colour light to dark green with dark mottling

green turtle

Threatened species and subspecies

Loggerhead turtle
Caretta caretta

Other names: Lauwora (Wunambal).

Description: The upper shell (carapace) has five pairs of costal scales (see p. 110).

Distribution and habitat: Loggerhead turtles inhabit coastal subtropical and tropical waters, with a temperature range of 16°–20°C worldwide, and are found in bays, coral reefs and estuaries. In WA, loggerhead turtles breed in Shark Bay (Dirk Hartog Island and Dorre Island), along the Ningaloo coast and adjacent beaches and in the Muiron Islands off North West Cape, with a few turtles nesting in the Dampier Archipelago.

Biology and ecology: Loggerhead turtles eat shellfish, crabs, sea urchins and jellyfish. The WA population migrates from feeding grounds in northern WA, the Northern Territory and probably Indonesia to breeding areas further south. Mating occurs from October to early December followed by nesting from late October to early March. Females lay up to six clutches per nesting season, with about 125 eggs in each clutch. Hatchlings emerge from late December to early April. Breeding and nesting occurs every two to five years. The sex of the hatchling is determined by incubation temperature, with warmer beaches producing predominantly female hatchlings.

Threats: Elsewhere in the world, loggerheads, like other turtles, have been killed in vast numbers for food. In WA, loggerheads have not been harvested to any significant extent, however, turtles that breed here may be killed elsewhere, particularly in south-east Asia. Loggerheads are threatened by foxes

predating eggs in the Ningaloo area. Vehicles driving on nesting beaches are also an issue. Loggerheads are captured in prawn trawls. In northern and eastern Australia, significant numbers of trawled turtles were drowned until 'turtle exclusion devices' (TEDs), also called 'bycatch reduction devices' or BRDs, recently became mandatory in trawls. In WA, TEDs are increasingly being used in prawn fisheries, greatly reducing the chance of turtle deaths. It is thought that longline fisheries drown many loggerheads, which take baited hooks. In the not-too-distant future, an increase in air and sea temperatures due to global warming could affect the species, by biasing the sex ratios of hatchlings.

Status: Endangered. The international status is the same. Loggerhead rookeries on Dirk Hartog Island and the Muiron Islands are being monitored. Fox predation of eggs on beaches in the Ningaloo area is a major concern. Some fox control is being undertaken, but more is required.

Above A female loggerhead turtle returns to the ocean after nesting.

Photo – Jiri Lochman

Below Each winter, baby loggerhead turtles are carried hundreds of kilometres south from northern breeding areas by means of the Leeuwin Current, and may be washed ashore on southern beaches.

Photo – Babs and Bert Wells /CALM

Above A green turtle at Ningaloo Marine Park.

Photo – Tony Howard

Green turtle *Chelonia mydas*

Other names: Juluwarru (Wunambal).

Description: The carapace of the green turtle has four pairs of costal scales, and the head has one pair of prefrontal scales (see diagram on p. 110). The high-domed shell is light to dark green with dark mottling.

Distribution and habitat: Green turtles have a worldwide distribution in tropical and subtropical seas with temperatures above 20°C. In WA, they nest on beaches from the Ningaloo coast northwards. There are significant rookeries on Barrow Island, in the Montebello Islands, the Dampier Archipelago and the Lacepede Islands, and small rookeries on many smaller Pilbara islands, as well as in the Kimberley.

Biology and ecology: Adults are herbivorous, feeding on seaweeds and seagrasses, while immature green turtles are carnivorous, feeding on jellyfish, small molluscs, crustaceans and sponges. Adult females breed only once every six years or so, although the nesting interval varies. The number of females breeding varies considerably from year to year, with very

little breeding occurring in some years and a lot in others. Females lay about 115 eggs per clutch and about five clutches per season. Hatchling sex ratios vary, depending on incubation temperature: warmer northern beaches produce predominantly females and southern beaches produce mainly males.

Threats: Green turtles have been drastically over-harvested in most parts of the world. In WA, a legal 'fishery' operated until the licences were cancelled and full legal protection was applied in 1973. Many turtles were taken from the North West Cape area and the Montebello Islands. High radiation levels resulting from an atomic weapon test in the Montebello Islands in October 1952 apparently killed large numbers of green turtles at Trimouille Island over the following summers. Hatchlings are disoriented by lights near the beach, including lights of yellow wavelengths such as low pressure sodium vapour lamps.

Status: Vulnerable. Their international status is Endangered. Green turtle populations have been studied at Barrow Island and in the Dampier Archipelago.

Hawksbill turtle *Eretmochelys imbricata*

Other names: Maral (Wunambal).

Description: The hawksbill turtle has a carapace with four pairs of costal scales. The head has two pairs of prefrontal scales (see diagram on p. 110).

Distribution and habitat: Hawksbill turtles forage near coral reefs in the warm tropical waters of the central Atlantic and Indo-Pacific regions. They rarely stray into temperate seas. In WA, they nest from the Ningaloo coast northwards, including the Lowendal Islands, the Dampier Archipelago and some other small Pilbara islands.

Biology and ecology: Hawksbill turtles feed on sponges, seagrasses, algae, sea squirts, soft corals and molluscs. Females migrate long distances between feeding and breeding grounds. They lay about 130 eggs per clutch, and hatchling sex ratios depend on incubation temperature.

Threats: Over-harvesting has been the cause of decline worldwide. Australia stopped trading in hawksbill turtle products in 1977, but in some parts of the world this species continues to be exploited for 'tortoiseshell' as well as for food.

Status: Hawksbill turtles are classed as Vulnerable in Western Australia, but their international status is Critically Endangered. The hawksbill is a highly endangered marine turtle on a worldwide basis, and WA possesses the only remaining large population in the entire Indian Ocean. A major colony at Rosemary Island in the Dampier Archipelago is being monitored by volunteers. Another nesting beach at Varanus Island was monitored for several years until the oil company employee doing the work moved on.

Below In many parts of the world, thousands of hawksbill turtles have been killed for the scales on their back, misleadingly named 'tortoiseshell', as well as for food. Despite international protection, the trade in tortoiseshell is continuing.

Photo – Jiri Lochman

Above An olive ridley turtle.

Photo – John McCann
/NatureFocus

Olive ridley turtle *Lepidochelys olivacea*

Other names: Pacific ridley.

Description: The carapace has six pairs of costal scales (see diagram on p. 110).

Distribution and habitat: Ridleys do not nest in WA, but data from the Department of Conservation and Land Management's WA Marine Turtle Research Project show that there is a small feeding population off the Kimberley. They live in shallow, protected tropical and subtropical seas throughout the world. Turtles found off WA probably breed in the Gulf of Carpentaria.

Biology and ecology: Olive ridleys eat jellyfish, starfish and small crabs.

Threats: Genetic research shows that the small Australian population is distinct from those of India and Malaysia in the Indian Ocean, and from those of Mexico and Costa Rica in the Pacific Ocean. Thus, although the species is more abundant in some other parts of the world, Australia is the custodian of an endemic stock. In the past, many have drowned in trawls in the northern Australian prawn fishery. The use of 'turtle exclusion devices' should greatly reduce this impact in the future.

Status: Endangered. Their international status is also Endangered.

Flatback turtle *Natator depressus*

Other names: Barwanjan (Bardi), madumal (Wunambal).

Description: The carapace has four pairs of costal scales. The head has one pair of prefrontal scales. The olive grey carapace has a low dome with upturned edges (see diagram on p. 110) and four pairs of costal scales.

Distribution and habitat: Flatback turtles breed only in Australia and are restricted to Australia and some nearby waters in Indonesia and Papua New Guinea. They inhabit coastal waters, rather than deep oceans.

Biology and ecology: Flatbacks feed on soft corals, jellyfish and other soft-bodied animals such as sea cucumbers. They nest on sheltered, often muddy mainland beaches in the Kimberley and Pilbara, and on a few islands, including the Dampier Archipelago, Barrow Island, Thevenard Island and the Muiron Islands. Most rookeries are small. About 54 eggs are laid in each clutch, and up to three clutches are laid per season.

Threats: Eggs are eaten by foxes and dingoes. Some nesting beaches, for example in Port Hedland, are subject to street lighting that disorients nesting females and hatchlings when they are moving towards the sea. Flatbacks are caught in trawls and are subject to indigenous harvest.

Status: Vulnerable. This species is not currently included in the IUCN Red List.

Below Flatback turtles can be recognised by the distinctive flat-domed shell with its upturned edges.

Photo – Jiri Lochman

Leatherback turtle
Dermochelys coriacea

Other names: Luth.

Description: Adults can weigh up to 500 kg and their carapaces can attain lengths of 160 cm, making this the largest species of turtle. The leathery carapace has no scales. It is black with light spots and five longitudinal ridges.

Distribution and habitat: Leatherback turtles have the widest distribution of any turtle, but do not nest in WA. The closest rookeries are in Papua New Guinea, Indonesia and Malaysia. The turtles are regularly found in our coastal waters as far south as Cape Naturaliste, and occasionally stray into the Southern Ocean as far east as Esperance.

Biology and ecology: Leatherback turtles feed mainly on jellyfish.

Threats: Entanglement in fishing gear, including lobster pot lines, is the main cause of death off the WA coast. They also take longline fishing hooks and drown. The species is rare and its breeding grounds in many parts of the world are under threat, so even a few deaths each year can have a significant impact.

Status: Vulnerable. The international status of this species is Critically Endangered.

Above A leatherback turtle found dead in the Jurien area.

Photo – Shane O'Donoghue

Below This leatherback turtle, entangled in fishing gear, was later released.

Photo – Steve Sturgeon

Western swamp tortoise *Pseudemydura umbrina*

Other names: Short-necked tortoise, western swamp turtle.

Description: This small freshwater tortoise has a carapace length of up to 145 mm. The upper shell (carapace) is yellowish-brown to almost black, sometimes with a maroon tinge. The lower shell is yellow to brown or occasionally black, often with black spots on a yellow background and black edges to the scales. Its short neck easily distinguishes the western swamp tortoise from the only other freshwater tortoise (or turtle) found in the south-west of WA; the oblong tortoise (*Chelodina oblonga*) has a neck equal to or longer than the length of its shell.

Distribution and habitat: Western swamp tortoises occur naturally only in the Swan Valley and adjacent areas, on the outskirts of Perth. They have been recorded from near Pearce Air Force base south to Perth Airport. They inhabit temporary swamps on clay or on sand over clay. Anecdotal evidence that they occurred near Pinjarra and near Mogumber has never been verified. Western swamp tortoises have recently been translocated to Mogumber Nature Reserve.

Biology and ecology: During winter and spring the tortoises live in swamps, eating the abundant aquatic invertebrates. During summer, they aestivate (sleep) in holes in the ground or under deep leaf litter. Three to five hard-shelled eggs are laid in an underground nest during early summer, and hatch the following winter. It takes 10 to 15 years for western swamp tortoises to reach maturity and breed.

Threats: Most habitat was cleared or drained many decades ago. Predation by foxes and a series of drought years (when there was too little water in swamps for successful breeding to take place) caused a serious decline in numbers in the 1970s and 1980s. The nature reserves lie in Perth's developing north-eastern corridor, and careful planning is required to ensure housing and other developments do not impact on the nature reserves. Fire is a threat at Twin Swamps and Mogumber nature reserves. At Ellen Brook Nature Reserve all tortoises aestivate underground in naturally-occurring tunnels, where they are unaffected by summer wildfires.

Status: Critically Endangered. There are less than 50 mature individuals in the wild. Most wild tortoises live in two nature reserves, at Ellen Brook and Twin Swamps, in the Upper Swan–Warbrook area. Research and conservation management have occurred since the 1960s. A recovery plan for the western swamp tortoise involves management of the nature reserves, population and environmental monitoring, captive breeding at Perth Zoo, translocations and public education. Funding for the recovery plan has come from the Department of Conservation and Land

Below Western swamp tortoise.

Photo – Babs and Bert Wells /CALM

Above The western swamp tortoise is the most endangered reptile in Western Australia.

Photo – Babs and Bert Wells /CALM

Management, Environment Australia (including Natural Heritage Trust funding), Perth Zoo, the World Wide Fund for Nature, the Western Australian Water Corporation and numerous other helpers, including some Perth companies, schools and international herpetological societies. Fox-proof fences surround both nature reserves and groundwater is pumped to supplement one swamp. Research into the ecology of the species and translocations is being carried out by Gerald Kuchling from The University of Western Australia's School of Animal Biology.

Translocations to Mogumber Nature Reserve commenced in 2000, but a very hot fire burnt out the entire area in December 2002. Monitoring of radio-tracked tortoises suggested that as many as 50% of the tortoises died, and more may have died after the fire because of elevated soil temperatures due to lack of shade. Significantly, all radio-tracked tortoises aestivating in artificial aestivating tunnels installed by the recovery team survived. More of these tunnels will be installed in fire-prone areas.

Above The Hermite Island worm lizard has not been seen on Hermite Island since 1952.

Photo – David Bettini

Hermite Island worm lizard *Aprasia rostrata rostrata*

Description: This legless lizard has a snout-vent length of up to 109 mm and a tail length of 72 mm. The snout is long and angular. The upper surface of the body is pale brown, with three darker brown lines on the nape and one on the tail, and there is a brown stripe on the side of the head. Recent examination of WA Museum collections found a specimen of this subspecies from North West Cape. The taxonomy of this group is still unresolved.

Distribution and habitat: The subspecies is known from only two specimens collected on Hermite Island in the Montebello Islands in 1952 and the North West Cape specimen. It has not been found on Hermite Island since 1952, despite searches.

Biology and ecology: Worm lizards spend most of their lives underground and are rarely seen.

Threats: Predation by cats and rats may have reduced, or even eliminated, this subspecies on Hermite Island.

Status: Vulnerable. Further searches are required now that cats and (probably) rats have been eradicated.

Yinnietharra rock dragon   *Ctenophorus yinnietharra*

Other names: Yinnietharra crevice dragon.

Description: The Yinnietharra rock dragon is closely related to the ornate rock dragon (*C. ornatus*) of the south-west of WA, but is reddish above and has broad black and pale bands on the tail, distinguishing it from other dragon lizards. It has a snout-vent length of up to about 115 mm, and a tail length of 200–270 mm. Like most dragon lizards, the male is larger, and has brighter colouration on the body than the female.

Distribution and habitat: As the name suggests, this species was first recorded on Yinnietharra Station in the Ashburton, where it is known from only a very small area. It lives on low granite rock outcrops interspersed with low shrubland.

Biology and ecology: Yinnietharra rock dragons shelter in crevices, or under exfoliating slabs on granite outcrops. They have not been studied, but probably eat insects and other invertebrates.

Threats: None known. This species is listed because of its extremely small geographic range. Removal for the illegal pet trade may be a threat.

Status: Vulnerable. The Department of Conservation and Land Management has carried out limited surveys of the area where it occurs. More thorough searches and research into conservation requirements are needed.

Above Male Yinnietharra rock dragons have a strongly-banded tail.

Photo – Ron Johnstone/ WA Museum

Below A female Yinnietharra rock dragon.

Photo – Andrew Burbidge /CALM

Airlie Island skink   *Ctenotus angusticeps*

Description: This small, slender, faintly patterned skink has a dark olive grey back, and sides that are mottled black and whitish. It has a snout-vent length of 59–74 mm, and the tail is 100–125 mm.

Distribution and habitat: This species was discovered on Airlie Island, a small 41 ha cay off the Pilbara coast. For some time it was not known from anywhere else, however, in 1990 a small population was located near the shore of Roebuck Bay, south of Broome. On Airlie Island, the skinks utilise all habitats: acacia shrubland, coastal spinifex (*Spinifex longifolius*) and tussock grass. Near Broome, they inhabit samphire flats.

Biology and ecology: Poorly known. Members of this genus are active, ground-dwelling skinks that eat insects and other invertebrates.

Threats: Airlie Island has been used as an oil storage base for a nearby undersea oilfield for some years. This does not seem to have significantly affected the skink, except to reduce the total available habitat and, consequently, the size of the population. The introduction of rats might be devastating.

Status: Vulnerable. The Airlie Island population has been monitored by the oil company.

Below The introduction of predators such as black rats could have devastating consequences for the Airlie Island skink.

Photo – Ron Johnstone /WA Museum

Lancelin Island skink
Ctenotus lancelini

Description: The Lancelin Island skink was originally described as a subspecies of the widespread and abundant red-legged skink (*Ctenotus labillardieri*). However, recent genetic work has indicated that it is sufficiently different from this mainland species to warrant full species status. This small skink (220 mm long and weighing around 10 g at maturity) has an elongate body and relatively short limbs for a skink in this genus, reflecting its habit of foraging on and under leaf litter. The upper body is pale brown, with several wavy lines of dark spots. The tail is grey with dark flecks, the flanks have several white stripes and the legs are yellow.

Distribution and habitat: The species is known only from the 7.6 ha Lancelin Island, offshore from Lancelin. A single skink was found on the adjacent mainland in 1994. Despite further survey for the species in this area, no more individuals have been captured, so the status of the skink on the mainland remains unclear. Lancelin Island skinks use all of the main vegetated habitat types available on the island, including sand, and shallow soil over limestone. The vegetation consists of low shrubs with a variable cover of winter annuals, predominantly introduced rye grass (*Lolium rigidum*) and wild oats (*Avena barbata*). Favoured areas, especially for breeding females, have nearby slopes facing north to north-east, which are protected from prevailing southerly winds and receive sun early in the day.

Biology and ecology: Lancelin Island skinks forage within leaf litter, where they eat insects and spiders. Females lay clutches of two to five eggs in early summer, and hatchlings emerge in mid to late summer. They are active by day and, if startled, will retreat to the burrows of nesting seabirds.

Above The Lancelin Island skink is the most geographically restricted vertebrate animal in Western Australia.

Photo – Ron Johnstone /WA Museum

Threats: The introduction of predators, including rats and mice, might have catastrophic consequences. Lancelin Island is popular for boating, surfing and windsurfing, and trampling of vegetation or the risk of fire are significant threats. Weed invasion and a now very large silver gull (*Larus novaehollandiae*) colony do not seem to be adversely affecting the skinks.

Status: Vulnerable. Lancelin Island is a nature reserve managed by the Department of Conservation and Land Management. In 1992, the Western Australian Society of Amateur Herpetologists reported that searches for the skink found only one individual, and expressed concern about the future of the species. The Department, with financial support from Environment Australia, responded by commissioning research undertaken by Barb Jones that showed that the skinks were abundant and, although highly restricted, not under immediate threat. A recovery plan was prepared and a recovery team appointed to implement the plan. A captive breeding colony was established at Perth Zoo, and captive-bred skinks were released on an island in Jurien Bay in March and December 2001. A number of translocated skinks have been recaptured during sampling trips, but no juveniles have yet been captured, so there is no indication that breeding is occurring.

Hamelin ctenotus   *Ctenotus zastictus*

Description: The Hamelin ctenotus has a snout-vent length of up to 60 mm and a tail length of up to 155 mm. This long-tailed, medium-sized skink has a blackish back and sides, with eight white stripes and four series of pale brown spots.

Distribution and habitat: The species is restricted to a small, isolated area of spinifex (*Triodia*), with open eucalypts on red sand, on Cockburn and Hamelin stations south of Shark Bay.

Biology and ecology: Not studied.

Threats: There are no current obvious threats. Habitat degradation through frequent burning or stock grazing are potential threats.

Status: Vulnerable. Field surveys to determine the distribution, habitat preferences and conservation status of the Hamelin ctenotus are required.

Below The strongly-marked Hamelin ctenotus is restricted to a small area of spinifex hummock grassland south of Shark Bay.

Photo – Greg Harold

Tjakurra *Egernia kintorei*

Other names: Giant desert skink. Tjakurra is the Pitjantjatjara and Ngaanyatjarra name for this species.

Description: The tjakurra has a snout-vent length of around 200 mm, and a tail length up to 240 mm, with a total length up to 440 mm. It weighs up to 350 g. The colouration of this large, robust, reddish skink distinguishes it from other *Egernia* species. The undersurface is creamy yellow, but turns bright yellow in the breeding season. The pupil of the eye is vertical (like that of a cat) and the skink is active in the late afternoon and early evening. The legs are powerfully built and used for vigorous digging.

Distribution and habitat: This skink is sparsely distributed in the Great Sandy, Gibson and northern Great Victoria deserts of WA. It also occurs in the south-western Northern Territory and in north-western South Australia. Tjakurra inhabit sandy and gravelly deserts dominated by spinifex (*Triodia*) hummock grassland.

Biology and ecology: Tjakurra live communally in large burrow systems, up to 10 m in diameter and 1 m deep, dug into sandy soils. Young adults leave their birth burrows in their third summer, and may move into a burrow vacated by other lizards or mulgara. Tjakurra burrows typically have a large latrine area outside the main burrow entrance, where resident lizards habitually defecate. Looking for burrows with these distinctive latrines is the easiest way to find tjakurra populations. Tjakurra are omnivorous, eating a wide range of invertebrates, especially termites, as well as succulent leaves, flowers and fruits.

Threats: Predation by foxes and cats may be the main cause of decline, but it appears that fire has an important and perhaps synergistic role. The skinks need mature spinifex for shelter, in close proximity to burnt areas that contain preferred food plants, so changed fire regimes (from small fires to large wildfires) reduce the availability of suitable habitat.

Status: Vulnerable. Tjakurra had not been recorded in WA between 1950 and 1997, when Pintupi women took Department of Conservation and Land Management Research Scientist David Pearson to a population in the Gibson Desert. Since that time, the local Aboriginal communities have worked with arid zone ecologist Steve McAlpin on a Threatened Species Network grant to survey for other populations in the Gibson Desert. A recovery plan was prepared by Steve McAlpin and a recovery team set up in 1999.

Above The reddish-brown upper parts of the tjakurra camouflage it on the red sands of the interior.

Photo – Greg Harold

Above Greyish-black Baudin Island spiny-tailed skinks shelter under limestone rock slabs.

Photo – Andrew Burbidge /CALM

Below right Baudin Island spiny-tailed skink.

Photo – Greg Harold

Baudin Island spiny-tailed skink   *Egernia stokesii aethiops*

Description: Compared with other spiny-tailed skinks, this subspecies is relatively small and has dark colouration with no patterning on the back or sides. The body is flattened and covered with strongly keeled scales. The broad, flat tail has large spines that help the skink to grip tightly in crevices. The taxonomy of the *Egernia stokesii* species group is being reviewed. Other isolated populations of *Egernia stokesii* have been located, and genetic work has commenced to clarify the relationships of the group, including those of the Baudin Island population.

Distribution and habitat: This subspecies is restricted to Baudin Island in Shark Bay.

Biology and ecology: The skinks shelter under limestone slabs. Little is known of their biology, but, like other *Egernia*, they probably eat insects and vegetation, such as leaves, fruit and flowers, and produce live young. Other larger *Egernia* are completely insectivorous as juveniles, becoming omnivorous as they reach adulthood. These skinks exhibit strong social bonds. Rock refuges often contain several individuals, including adults with large young.

Threats: There are no obvious current threats. Potential threats include the introduction of predators to the island.

Status: Vulnerable. Baudin Island is a nature reserve.

Western spiny-tailed skink *Egernia stokesii badia*

Description: This species has a snout-vent length of 81–194 mm, a tail up to 85 mm, and a total length of up to 275 mm. This large, robust, spiny-tailed skink has a brown or reddish-brown back marked with angular greyish-white spots. The taxonomy of the *Egernia stokesii* group is currently being reviewed.

Distribution and habitat: The western spiny-tailed skink is found in the central Wheatbelt, with two outlying records inland from Carnarvon. In cleared areas, it has declined significantly.

Biology and ecology: Limited information suggests that western spiny-tailed skinks live in small groups, shelter in hollow logs and rocky outcrops, and are active during the day. Animals deposit faecal droppings just outside the refuge, behaviour shared with other large *Egernia* species. They produce live young.

Threats: Habitat clearing for agriculture has been the major cause of decline. Predation by foxes is a continuing threat. A survey by WA Museum staff in 1998, funded by the BankWest *LANDSCOPE* Conservation Visa Card, showed that the subspecies survives in reserves with good quality York gum (*Eucalyptus loxophleba*) woodland and in similar country east of cleared farmland. As a result, the status of the species was changed from Endangered to Vulnerable.

Status: Vulnerable.

Above An inhabitant of woodlands, the western spiny-tailed skink has suffered a large range reduction due to land clearing.

Photo – Ron Johnstone /WA Museum

Buccaneer burrowing skink *Lerista praefrontalis*

Description: This small *Lerista* has no trace of forelimbs, two toes on the small hind feet and a movable eyelid. Its upper surface is pale reddish-brown, with darker markings. Unlike any of its close relatives, this species possesses prefrontal scales on the head.

Distribution and habitat: A single specimen was collected on King Hall Island, in the Buccaneer Archipelago north of Derby, during a Department of Fisheries and Wildlife expedition in 1982. A 1992 search of King Hall Island located only *Lerista griffini*, and it has been suggested that the type specimen may be an aberrant member of this species. Further taxonomic research, combined with field survey, is required.

Biology and ecology: The skink was found in litter and sand at the base of a cliff.

Threats: There are no obvious current threats. Potential threats include the introduction of predators, especially rats.

Status: Vulnerable.

Above This Pilbara olive python has recently eaten a fairly large animal.

Photo – Ron Johnstone /WA Museum

Pilbara olive python *Morelia olivacea barroni*

Description: The largest snake in WA, the Pilbara olive python has been recorded to lengths of 4.5 m with weights up to 15 kg. Reports of larger specimens are unreliable due to the difficulties of measuring large snakes and the lack of specimens. This snake has a dull olive or reddish-brown upper surface, and is pale cream below. Recent research by Lesley Rawlings suggests that the Pilbara olive python may be a full species, rather than a subspecies of the olive python found in the Kimberley, northern Northern Territory and western Queensland.

Distribution and habitat: The Pilbara olive python is restricted to the Pilbara and northern Ashburton region, where it inhabits moist areas such as gorges, rivers, pools and surrounding hills.

Biology and ecology: Our knowledge of this python's ecology is almost entirely due to the efforts of volunteers and Department of Conservation and Land Management rangers, who have participated in a radio-telemetry study at several sites in the Pilbara. These pythons are usually found close to water and rock outcrops. They have discrete home ranges

and shelter in logs, flood debris, caves, tree hollows and thick vegetation. Prey includes euros, rock-wallabies, flying-foxes, ducks and pigeons, which they ambush on animal trails or at waterholes. During June and July, males may travel long distances to locate females for breeding. They remain together for up to three weeks, probably mating a number of times. The male then returns to his usual home range. In November, breeding females lay eggs, which hatch about two months later in mid-January. The juveniles disperse widely and often appear in towns such as Tom Price, attempting to raid cages for small birds.

Threats: Habitat destruction for industrial development, accidental or deliberate killing on roads, around houses and near swimming places, predation of juveniles by foxes and feral cats, and large summer wildfires are all threatening this python.

Status: Vulnerable. Research into the basic ecology of animals at Millstream-Chichester National Park and on the Burrup Peninsula with the Nickol Bay Naturalist Club is ongoing.

References

Aplin, K.P. and Smith, L.A. (2001). Checklist of the frogs and reptiles of Western Australia. *Records of the Western Australian Museum* Supplement No. 63, 51–74.

Burbidge, A.A. and Kuchling, G. (2003). Western swamp tortoise recovery plan. 3rd edition January 2003 to December 2007. Department of Conservation and Land Management, Bentley.

Cogger, H.G., Cameron, E.E., Sadlier, R.A. and Eggler, P. (1993). *The action plan for Australian reptiles.* Australian Nature Conservation Agency, Canberra.

Environment Australia (1998). *Draft recovery plan for marine turtles in Australia.* Environment Australia, Canberra.

How, R.A., Dell, J. and Aplin, K.P. (1999). Assessment of the central wheatbelt populations of the endangered skink *Egernia stokesii badia.* Western Australian Museum of Natural Science, Perth.

Pearson, D. and Jones, B. (2000). Lancelin Island skink recovery plan. Wildlife Management Program No. 22. Department of Conservation and Land Management, Bentley.

Maryan, B. and Robinson, D. (1997). An insular population of *Lerista griffini* and comments on the identity of *Lerista praefrontalis* (Lacertilia: Scincidae). *Western Australian Naturalist* 21, 157–160.

Storr, G.M., Smith, L.A. and Johnstone, R.E. (1983). *Lizards of Western Australia. II. Dragons and monitors.* Western Australian Museum, Perth.

Storr, G.M., Smith, L.A. and Johnstone, R.E. (1990). *Lizards of Western Australia. III. Geckos and pygopods.* Western Australian Museum, Perth.

Storr, G.M., Smith, L.A. and Johnstone, R.E. (1999). *Lizards of Western Australia. I. Skinks.* Revised edition. Western Australian Museum, Perth.

Storr, G.M., Smith, L.A. and Johnstone, R.E. (2002). *Snakes of Western Australia.* Revised edition. Western Australian Museum, Perth.

Turtle Conservation Fund (2002). *A global action plan for the conservation of tortoises and freshwater turtles. Strategy and funding prospectus 2002–2007.* Conservation International and Chelonian Research Foundation, Washington, DC.

Below A green turtle at Ningaloo Marine Park.

Photo – Tony Howard

Amphibians

chapter 7

Above The threatened orange-bellied frog.

Opposite Because of the unusual colouring of its underparts, the Vulnerable sunset frog is one of the world's most recognisable frogs.

Photos – Grant Wardell-Johnson

Worldwide, there are three major groups of amphibians: frogs; salamanders and newts; and legless amphibians (caecilians), sometimes known as 'rubber eels'. Only frogs occur in Australia. Considering its aridity, Australia has a relatively large number of frog species, including many that are specially adapted to living in dry places. Slightly more than 210 species have been described, but some experts think that the real total may be as high as 250.

Australian frogs belong to four families: about 75 species of tree frogs or Hylidae (Pelodryadidae according to some authors); about 120 species of ground-dwelling and burrowing frogs or Myobatrachidae (Leptodactylidae according to some authors); about 18 species of narrow-mouthed frogs in the Microhylidae; and one species of true, or old world, frog in the Ranidae. Of these, only Hylidae and Myobatrachidae are found in Western Australia. Fifteen of the 30 genera of Australian frogs occur here. There are 78 frog species in WA, about a third of the total frog fauna of the continent. Some 38 frog species are confined to WA.

Amphibians are particularly susceptible to environmental change. Because they readily absorb chemicals through the skin, frogs are often used as indicators of environmental health. In Australia, four frogs are known to have become extinct and 26 species are listed as threatened. Worldwide there have been seven extinctions and 157 species are threatened. Thus, Australia has more than 50% of the world's recently extinct amphibian species and more than 15% of threatened species.

Despite major declines of frogs in many parts of the world (including eastern Australia), WA species have fared relatively well so far, with no extinct species. Only three species are listed as threatened, one of which is Critically Endangered, while the other two are Vulnerable. However, not all unlisted frogs are doing well—there has been extensive loss of habitat in cleared areas,

especially in the Wheatbelt, where most wetlands have either disappeared or become saline.

The recently-discovered chytrid fungus (*Batrachochytrium dendrobatidis*), thought to have been introduced from Africa or South America, causes a disease known as chytridiomycosis in frogs. It is thought to have caused the extinction of some frog species in eastern Australia. The fungus is present in many species of frogs in the south-west of WA, including some threatened species, but so far does not seem to be causing species to decline significantly. Further study and population monitoring are needed.

Another threat that is on its way to WA is the cane toad (*Bufo marinus*). Cane toads were introduced to Queensland from Hawaii (they are actually a native of the Americas, originally occurring from Texas south to the Amazon basin) in 1935, to control beetle pests in sugar cane. Instead of controlling insect pests, the cane toad itself became a pest, rapidly spreading north and south. It is currently moving westward, through the Northern Territory, at 30 to 35 kilometres each year. By 2004 it was close to the WA border. It will doubtless establish throughout much of the Kimberley, where it can be expected to affect some native animals, particularly predators such as the northern quoll (*Dasyurus hallucatus*), via its highly toxic skin glands. Its effects on native frogs are unclear but, based on information from Queensland and the Northern Territory, are likely to be minor.

Threatened species and subspecies

White-bellied frog   *Geocrinia alba*

Description: This very small frog has a snout-vent length of just 17–24 mm. The back is light brown or grey, with darker brown spots, and the underparts are white.

Distribution and habitat: The white-bellied frog occurs within a restricted range in the Witchcliffe-Karridale area, south of Margaret River, where it has been located at 56 sites. The range of this species is 101 km², but the actual area it occupies is only about 193 ha. The frogs live in swampy valley bottoms adjacent to creeklines.

Biology and ecology: The males call, in spring to early summer, from small depressions in moist soil under litter and dense vegetation. The call is a discrete train of 11–18 pulses repeated rapidly, almost too fast to be resolved by the human ear. The eggs are surrounded by a mass of jelly and deposited in small depressions. After hatching, the tadpoles develop with no free-swimming or feeding stage. Unlike most frogs, individuals move over very short distances and populations are genetically isolated from each other.

Threats: Many white-bellied frog populations are on private property used for cattle grazing, and, increasingly, for viticulture or eucalypt plantations. The frog populations can survive if the vegetation fringing the creekline is left undisturbed. However, damage by cattle and fire, and alteration of drainage patterns or local hydrology (for example, by damming creeks for irrigation of vines or through bluegum plantation development) causes the decline and disappearance of populations, with little chance of natural reinvasion.

Status: Critically Endangered. A recovery plan is being implemented by a recovery team. Much research has been conducted by the School of Animal Biology at The University of Western Australia. Frog populations are being monitored by Department of Conservation and Land Management staff. The recovery team has provided funding for some landholders to fence their creeks. The purchase in 2000 of a significant area of land that will link the proposed Forest Grove and Blackwood River national parks, funded jointly by the Commonwealth and State governments, has ensured protection for a number of significant white-bellied frog populations.

Below A male white-bellied frog in its breeding burrow. The species was discovered as recently as 1983 by frog researchers Grant Wardell-Johnson and Dale Roberts in the jarrah forest of the Leeuwin-Naturaliste National Park.

Photo – Grant Wardell-Johnson

Orange-bellied frog   *Geocrinia vitellina*

Description: The orange-bellied frog is very similar to the white-bellied frog, but it has paler upper parts and the front half of the underparts is orange.

Distribution and habitat: The orange-bellied frog occurs in a very small area east of Witchcliffe, where its range is 6.3 km² but the actual area it occupies is only 20 ha, the smallest of any mainland vertebrate animal in Australia. Its habitat is similar to that of the white-bellied frog. It occurs within broad, U-shaped valleys in the lower reaches of six creeklines that drain into the Blackwood River.

Biology and ecology: Its biology and ecology are similar to those of the white-bellied frog. The call is a discrete train of 9–15 pulses, repeated just slowly enough to be resolved by the human ear.

Threats: Frequent or hot summer fires are a major threat, although habitat damage by pigs would be of concern should pig numbers increase in this area.

Status: Vulnerable. The entire range of this species lies within State forest, which is a proposed national park. A recovery plan is being implemented by a recovery team. The School of Animal Biology of The University of Western Australia has conducted considerable research into this species. The frog populations are being monitored by Department of Conservation and Land Management staff. Pig control is being undertaken by local people.

Above Though the sunset frog was first found in 1994, it is now known from 24 populations and its status has been altered from Endangered to Vulnerable.

Photo – Grant Wardell-Johnson

Sunset frog   *Spicospina flammocaerulea*

Other names: Harlequin frog, mountain road frog.

Description: The sunset frog is a moderate-sized species (the snout-vent length of females is 31–36 mm and that of males is 29.5–34.8 mm) characterised by massive glands between its prominent eyes and front legs. The front half of its lower body is orange and the rear half is covered with striking light blue spots on a dark grey to black background.

Distribution and habitat: The sunset frog is restricted to a small area east and north-east of Walpole, near and north of Bow Bridge. It inhabits peat-based swamps in the headwaters of streams or perched swamps in areas of diffuse drainage. Most known sites are in and around the northern and eastern edge of the tingle forests. Populations have been found in the Bow River, Kent River and Frankland River drainages.

The known 'extent of occurrence' is now 305 km², while the 'area of occupancy' is about 120 ha. The species is known from at least 24 locations, of which 12 are on private property. Ten locations are in national parks or State forest proposed for reservation as national park.

Biology and ecology: Sunset frogs are conventional aquatic breeders and the eggs are deposited singly in shallow, still and slow-flowing water. Males have been heard calling from September to December, but peak activity seems to be in November and early December at most sites. The call, formed of two pulsed notes, is unlike any other species in south-western Australia. Nothing is known of food preferences. Marked animals have been recaptured two years after initial marking, suggesting that this species may have a relatively long life span and fairly good adult survival.

Threats: Recent research suggests that the species is not significantly threatened. For example, sunset frogs have now been found in several highly modified swamps within farmland, and the species calls more commonly after swamps have been burnt. Eight localities are within areas that are likely to be flooded or have their surface and groundwater systems significantly altered by the planned Bow River Dam. Possible threatening processes include inappropriate fire regimes; physical damage to swamps such as breaching of peat; damage by feral pigs; siltation from poorly-designed or executed road construction; loss of swamp vegetation due to dieback, resulting in open swamps lacking cover; construction of dams and consequent flooding or degradation of habitat; impacts of possible mining activity (exploration or development as per the Regional Forest Agreement) in State forest; pollution of swamps, for example, by chemicals used on farms; and illegal collection of the species due to its attractiveness and apparent rarity.

Status: Vulnerable. Sunset frogs were discovered in 1994. In 1997, the species was known from only four localities and had a known extent of occurrence of 3.63 km². At that time the species was considered to be Endangered. Thanks to research carried out by Dale Roberts and his colleagues from The University of Western Australia's School of Animal Biology we now know that there are at least 24 populations and that it occurs over a much wider area. This research also investigated survey methods and the biology and ecology of the species. A recovery plan is being implemented by a recovery team.

Below The sunset frog inhabits peat-based swamps.

Photo – Grant Wardell-Johnson

References

Aplin, K. and Kirkpatrick, P. (2001). 'Putting the finger on the frog fungus'. *LANDSCOPE* 16(3), 10-16.

Burbidge, A.A. and Roberts, J.D. (2002). Sunset Frog Recovery Plan. Wildlife Management Program No. 35. Department of Conservation and Land Management, Como.

Tyler, M.J. (1997). *The action plan for Australian frogs.* Wildlife Australia, Canberra.

Tyler, M.J., Smith, L.A. and Johnstone, R.E. (2000). *Frogs of Western Australia.* Western Australian Museum, Perth.

Wardell-Johnson, G., Roberts, J.D., Driscoll, D. and Williams, K. (1995). Orange-bellied and white-bellied frogs recovery plan, 2nd edition. Wildlife Management Program No. 19. Department of Conservation and Land Management, Como.

Fish

Fish is a collective term for four major groups of vertebrates: lampreys and hag fish (Cephalaspidomorphi); sharks, skates, rays and chimaeras (Elasmobranchii); bony fish (Actinopterygii); and a group of primitive fish known as coelacanths (Sarcopterygii or Crossopterygii) thought to have become extinct 65 million years ago until rediscovered in 1938. Today a small number of species of coelacanths, perhaps only two, exist in deep oceans off eastern Africa, Madagascar and Sulawesi in Indonesia. Only the first three groups occur in Australia.

Fish occur in marine, estuarine and freshwater habitats, with some species able to tolerate a wide range of salinities and temperatures and others unable to exist outside specialised habitats. Fish have many more species than other vertebrate groups. Worldwide, there are more than 20,000 species, while in Australia there are more than 1000 species of fish in southern Australian seas and many more in tropical oceans. In February 2003, the Western Australian Museum's FaunaList included 3207 species of fish in 293 families.

Threatening processes for fish include over-exploitation and habitat destruction, especially of estuaries, rivers and lakes near urban areas, and deteriorating water quality. Damage to coral reefs, mangroves and other specialised marine habitats is also a major problem. Most of the world's threatened fish are freshwater bony fish, but there is increasing concern about the status of sharks and rays, as commercial fishing, mainly outside Australia, is killing them in vast numbers, both directly and indirectly through bycatch. Lampreys, which breed in the headwaters of rivers and migrate to and from the ocean, are under increasing pressure because of obstructions to migration routes in rivers by dams, gauging weirs and even road culverts.

In WA, we are fortunate in having largely undamaged marine habitats. However, many of our estuaries and rivers are under considerable pressure, particularly from eutrophication due to increased nutrient levels derived from fertilisers applied to farms, salinisation of inland waterways and acid-sulphate soils. Rising levels of fishing and shell collecting, and increased boat ownership, mean that we need to ensure that aquatic habitats are conserved, including protection in a comprehensive, adequate and representative system of marine conservation reserves, and we need to ensure that takes of exploited species are managed in a sustainable manner.

Management of exploited fish species is the responsibility of the Department of Fisheries, which implements the Fish Resources Management Act. The creation of marine conservation reserves is included in the Conservation and Land Management Act, administered by the Department of Conservation and Land Management, with the reserves being vested in the Marine Parks and Reserves Authority.

No fish species has become extinct in WA. Two species of sharks are listed as threatened because they have been over-exploited, and more species in this group may be listed in future. Of particular concern are tropical sawfish and whiprays, which are killed as bycatch during net fishing, for example, for barramundi (*Lates calcifer*), as well as being hunted for trophies.

While fishing is a major commercial and recreational pursuit, no species of bony fish is currently threatened with extinction, except for two specialised fish that dwell in caves on the North West Cape peninsula. However, some species

have been depleted sufficiently to require special management restrictions. Over-collecting of some species for the aquarium trade is a concern, with the leafy seadragon (*Phycodurus eques*) included in the Department of Conservation and Land Management's List of Priority Fauna under Priority 2 (see Chapter 1).

There are relatively few species of freshwater bony fish in the south-west of WA. This is because we have short river systems, which often stop flowing during the dry summer and autumn months. Nevertheless, the south-west has several endemic species, most with relatively small geographic ranges, some of which have a highly specialised biology. Of particular interest is the salamanderfish (*Lepidogalaxias salamandroides*), which is largely restricted to a small area of coastal peat flats and adjacent forested areas between Walpole and Windy Harbour, with outlying populations north to Margaret River and east to Albany. It inhabits dark, acidic water in shallow pools and survives drying of this habitat, either by aestivating or sheltering in moisture below the surface, reappearing after autumn rains.

While no south-west freshwater fish species is listed as threatened, many have undergone significant reductions in their ranges, due to increasing salinity, land clearing and river and creek damming. These include: the pouched lamprey (*Geotria australis*), listed as Priority 1, which is declining because of river damming; the trout minnow or spotted minnow (*Galaxias truttaceus*), a Priority 1 species, which in WA has been reduced to two small creeks; the black-stripe minnow (*Galaxiella nigrostriata*), Priority 3; Balston's pygmy perch (*Nannatherina balstoni*), listed as a Priority 4 species; and the mud minnow (*Galaxiella munda*), listed as Priority 4.

In arid parts of the State, freshwater fish occur in some of the major river systems that retain pools of water year-round, such as the Fortescue. One arid zone species, the Fortescue grunter (*Leiopotherapon aheneus*), is included in the Priority Fauna List as Priority 4.

There are many more freshwater fish species in the Kimberley, with some found only in one or a few drainages. Some with very restricted distributions are included in the Department of Conservation and Land Management's Fauna Priority List and require monitoring.

Fish introductions, discussed in Chapter 2, could lead to native species becoming threatened. Of concern in WA are trout (*Salmo trutta* and *Oncorhynchus mykiss*), redfin perch (*Perca fluviatilis*), guppies (*Poecilia reticulata*), tilapia (*Tilapia mossambicus*) and gambusia (*Gambusia holbrooki*). Exotic aquarium fish are sometimes released into the wild, and some of these have the potential to become pests, for example, carp (*Cyprinus carpio*), goldfish (*Carassius auratus*) and swordtails (*Xiphophorus helleri*).

Threatened species

Grey nurse shark *Carcharias taurus*

Other names: Known as the sand tiger shark in the USA and the spotted ragged-tooth shark in South Africa.

Description: Grey nurse sharks have a large, stout body and are grey to greyish-brown above and off-white below. Reddish or brownish spots may occur on the tail fin and the posterior half of the body. They grow to at least 3.6 m. They are slow but strong swimmers, and are thought to be more active at night.

Distribution and habitat: The grey nurse shark is a cosmopolitan species found in inshore subtropical and temperate waters around the main continental landmasses, except in the eastern Pacific Ocean off North and South America. It generally inhabits inshore rocky reefs. In Australia, it has been regularly recorded from Mooloolaba in Queensland around most of the southern half of the continent to Shark Bay in WA, but is uncommon in Victoria, Tasmania and South Australia and absent from the Great Australian Bight. It has occasionally been recorded off the North West Shelf.

Grey nurse sharks are often seen hovering almost motionless just above the seabed, in or near deep sandy gutters or rocky caves, and in the vicinity of inshore rocky reefs and islands. They are usually found at depths of between 15 m and 40 m.

Biology and ecology: Adults eat a wide range of finfish, other sharks and rays, squid, crabs and lobsters. Both sexes mature at about 2.2 m, when they are 4–6 years old. In captivity they have lived to 16 years, however, in the wild their life span is probably less than this.

Threats: Grey nurse sharks have been greatly depleted through commercial shark fishing, amateur line fishing and, particularly, spearfishing. Despite their fearsome appearance they are harmless to people.

Status: Vulnerable. The east coast population, which occurs mainly in New South Wales, is listed as Critically Endangered. The species' worldwide status is Vulnerable. A national recovery plan has been prepared by Environment Australia.

Below A grey nurse shark with attendant remora.

Photo – Jiri Lochman

Above Great white shark.

Photo – Ron and
Valerie Taylor/NatureFocus

Great white shark *Carcharodon carcharias*

Other names: White pointer, white shark.

Description: Great white sharks have a moderately stout, torpedo-shaped body and are grey to greyish-brown on their upper surface and white below. They grow to at least 6 m in length, with unconfirmed reports of sharks up to 7 m.

Distribution and habitat: The great white shark is distributed widely in temperate and subtropical oceans throughout the northern and southern hemispheres, though it prefers temperate waters. It is more frequently encountered in South Africa, California, the north-east coast of the USA and southern Australia, usually around rocky reefs and islands in inshore coastal waters. In Australia, it is found from Moreton Bay in Queensland to the North West Cape in WA, but is more frequently seen in the south, particularly near seal and sea-lion colonies. Great white sharks are uncommon.

Biology and ecology: Large adults feed mainly on seals, whales and dolphins, but also take finfish and other sharks. Juveniles feed mainly on fish. Great white sharks occasionally attack people, with several deaths being attributed to them over the past 100 years or so. However, as great white sharks were once quite common in coastal Australian waters, and still occur in lower numbers, they clearly do not seek out human prey. Being top predators, however, they are inquisitive and unafraid, and will approach small boats and even bite outboard motors. Females mature at 4.5–5 m and attain greater lengths and weights than males, which mature at about 3.9 m. Minimum ages at maturity have been estimated to be 11 years for females and nine years for males.

Threats: Commercial and amateur fishing, including bycatch in longlines and nets, threatens this species. Game fishing was a threat until the species was listed as threatened and legally protected. In 1997, it was estimated that around 500 great white shark mortalities in Australian waters each year may have been due to human activities, of which 300 were most likely related to commercial fishing including capture as bycatch. Although the total Australian great white shark population is not known, it has clearly declined significantly and is thought to be too small to sustain such a high level of mortality in the long term.

Status: Vulnerable. A draft national recovery plan has been developed by Environment Australia.

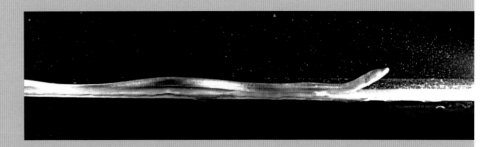

Blind cave eel   *Ophisternon candidum*

Description: The blind cave eel is long and slender, white or pink in colour and can reach up to 40 cm long. This fish has no fins apart from a rayless membrane around the tip of the tail.

Distribution and habitat: The species occurs only on coastal plains on the North West Cape peninsula, where it has been recorded in 11 caves. Where present it is rare.

Biology and ecology: The blind cave eel is a carnivore that feeds on aquatic invertebrates living in groundwater (stygofauna). It lays eggs.

Threats: Groundwater pollution and lowering of groundwater tables, through extraction for human use, are the main threats.

Status: Vulnerable. The Department of Conservation and Land Management's North West Cape Karst Management Advisory Committee is the recovery team for this species and the blind gudgeon.

Above Blind cave eel.
Photo – Geoff Taylor
/Lochman Transparencies

Below Blind cave eel.
Subterranean fish may appear pink because their blood is visible through their colourless skin.

Photos – Douglas Elford
/WA Museum

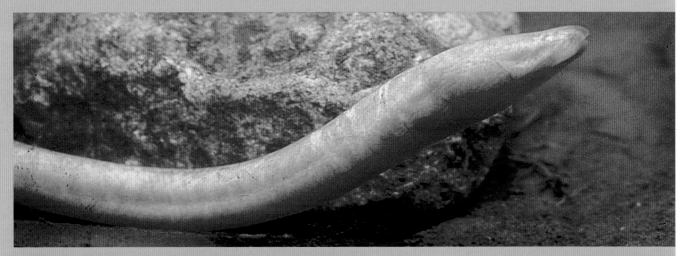

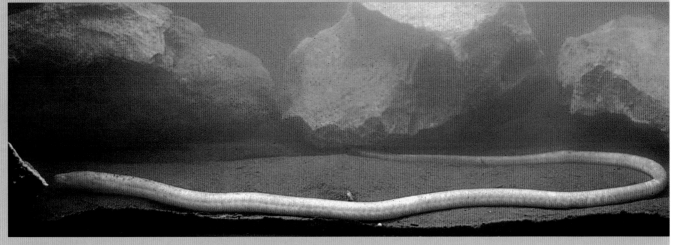

Above and below The blind gudgeon inhabits groundwater on the North West Cape peninsula and Barrow Island.

Photos – Douglas Elford /WA Museum

Blind gudgeon   *Milyeringa veritas*

Other names: Cave gudgeon.

Description: Blind gudgeons are about 4.5 cm long, and they are white to transparent. The head is completely scaleless, with several rows of small sensory papilla particularly obvious on the chin. Eyes are absent.

Distribution and habitat: The blind gudgeon occurs in subterranean waters on the coastal plains of North West Cape peninsula and on Barrow Island. It inhabits anchialine systems, where there is fresh or brackish water overlying underground seawater affected by marine tides. On North West Cape, it has been recorded in 24 caves up to 4.5 km inland. The species can be locally common.

Biology and ecology: Genetic data show that there are several isolated populations on the North West Cape peninsula. The blind gudgeon feeds on algae and small invertebrates living in groundwater.

Threats: Groundwater pollution and lowering of groundwater tables through extraction for human use are the main threats. Some caves on North West Cape, in which the species was once recorded, have been destroyed through filling, drying or siltation.

Status: Vulnerable. The Department of Conservation and Land Management's North West Cape Karst Management Advisory Committee is the recovery team for this species and the blind cave eel.

References

Environment Australia (2000). Draft recovery plan for Great White Sharks *Carcharodon carcharius* in Australia. Environment Australia, Canberra.

Environment Australia (2002). Recovery plan for the Grey Nurse Shark *Carcharius taurus* in Australia. Environment Australia, Canberra.

Humphreys, W.F. (1999). The distribution of Australian cave fishes. *Records of the Western Australian Museum* 19, 469–472.

Morgan, D., Gill, H. and Potter, I. (1996). The distribution of freshwater fish in the south-western corner of Australia. Water Resource Technical Series, Waters and Rivers Commission Report WRT4 1996. Waters and Rivers Commission, Perth.

Pollard, D.A. (Ed.) (1989). Introduced and translocated fishes and their ecological effects. Proceedings No. 8, Australian Society for Fish Biology Workshop. Australian Government Publishing Service, Canberra.

Wager, R. and Jackson, P. (1993). *The action plan for Australian freshwater fishes.* Australian Nature Conservation Agency, Canberra.

Invertebrates

Invertebrates is a collective term for animals without backbones, that embraces a wide array of animals as diverse as jellyfish, worms, crustaceans, insects, sea stars and molluscs. They occur in almost every terrestrial, marine and freshwater habitat, apart from the extreme polar regions and very high mountain peaks.

Invertebrates comprise more than 95% of all living species. There are between 40,000 and 50,000 vertebrate species in the world, whereas about 1.4 million species of invertebrates have been described. But there are many more undescribed invertebrate species—some estimates are as high as 50 million, although most are in the order of 10 to 30 million. Every year about 15,000 to 20,000 new invertebrate species are described.

Australia has about 6000 species of vertebrate animals and about 100,000 described species of invertebrate animals. There are, perhaps, 200,000 or more undescribed species. Most Australian invertebrates occur nowhere else.

Although invertebrates make up most biodiversity at the species level, they have received relatively little conservation attention. This is partly because of a lack of information—taxonomic, biological, ecological and conservation status—and partly because many people do not empathise with invertebrates to the extent that they do with vertebrates like mammals and birds. Scientists and naturalists sometimes describe larger animals as 'charismatic megafauna', because many people express concern only about them. Far fewer resources are allocated to study and conserve invertebrates. However, as more studies of ecosystem function become available, it is becoming increasingly clear that invertebrates are the key components of most ecosystems. Termites, for example, are a vital part of many arid ecosystems in Australia, where they recycle nutrients and in some ways fill a similar ecological role to large ungulates (such as antelopes) in African grasslands.

In 2001, six invertebrate species were listed as extinct Australia-wide, while 98 species were listed as threatened. Both numbers should be much higher—lack of knowledge has limited listing. In Western Australia, in 2003, six invertebrate species were listed as extinct and 77 as threatened. Many more species were listed in WA between 2001 and 2003, so comparisons of the proportion of WA-listed and Australian-listed species can not be made.

Threatening processes for invertebrates are similar to those for other animals. Land clearing is the major threat, with habitat destruction the main reason for the loss of invertebrates. Land degradation is another problem—dramatic changes to Rottnest Island's vegetation since European settlement has probably caused the extinction of the Rottnest bee (*Hesperocolletes douglasi*). Secondary salination resulting from clearing is still dramatically changing habitats in the Wheatbelt, and the Department of Conservation and Land Management's biogeographic survey of the Wheatbelt, part of the State Salinity Strategy, is predicting the extinction of hundreds of species of freshwater and terrestrial invertebrates if salinity is not controlled.

There are few studies of the effects of invasive invertebrates on native wildlife, but these introductions are probably a significant factor. Two native snail species from the south-west are thought to have become extinct because of the introduction of the predatory snail *Oxychilus* sp. from Europe. Invasive 'tramp ants' are eliminating native ants from metropolitan Perth and may be doing the same in some bushland areas. Introduced earthworms have replaced native species in some areas. The potential for further invasive species, such as fire ants, to affect native wildlife is immense.

Above In the last decade, the graceful sunmoth has been found at only three localities near Perth.

Photo – Terry Houston /WA Museum

Opposite The Endangered tree-stem trapdoor spider and its characteristic nest.

Photo – Barbara York Main
Inset illustration – Brad Durrant

Above The Bundera Sinkhole is part of a significant karst landscape, in which caves and other underground cavities readily develop because of the presence of soluble rocks. These features are frequently home to threatened invertebrates. In the case of the Bundera Sinkhole, seven species occur here and nowhere else.

Photo – Peter Kendrick /CALM

Subterranean invertebrates

Animals that live underground and which are totally adapted to life in the dark are called troglobites. They lack functional eyes and have usually lost all their body pigment, appearing white or translucent. The terms troglobite or troglofauna are usually restricted to terrestrial animals that live in air spaces (such as caves). Stygofauna are aquatic troglobites living in groundwater.

WA has large areas of karst–terrain with special landforms and drainage characteristics due to the greater solubility of certain rocks (notably carbonate rocks such as limestone, dolomite or magnesite) in natural waters. One of these, the Nullarbor region, is the largest karst area in the world. Other significant karsts occur on the North West Cape peninsula (including Cape Range), at Barrow Island, at Millstream in the Pilbara and in the southern and eastern Kimberley, where ancient coral reefs now outcrop as Upper Devonian limestone ranges—the Napier, Oscar and associated ranges and the Ningbing Range. In karst landscapes, caves and other underground cavities readily develop due to water infiltration and movement. Another frequent feature is the large number of localised species of terrestrial molluscs.

There are also significant areas of limestone and calcrete in ancient riverbeds, mostly in the arid interior, that flowed when the climate was much wetter than it is now. Recent research suggests that many ancient rivers (palaeorivers) flowed in valleys carved out by glaciers in Permian times, between 290 and 248 million years ago. Research at the WA Museum has demonstrated that groundwater (whether fresh, brackish or saline) within these features supports rich communities of stygofauna, and that communities in different palaeorivers may be very different from each other. Stygofauna is also found in groundwater beneath rivers that still flow after heavy rain, for example in the Pilbara, and in groundwater and surface water in caves in the south-west.

While research into the State's subterranean fauna is in its early stages, enough is known about a few of the communities and species for them to have been listed as threatened. Communities that are considered to be threatened with extinction are discussed in Chapter 10.

Conservation of invertebrates

At present, comparatively few resources are directed towards invertebrate conservation. Unless society directs more resources towards the study of invertebrate animals, including studies of their taxonomy, we will probably continue to lose many species without them even being lodged in a Museum collection and being described and named.

Extinct species

Rottnest bee *Hesperocolletes douglasi*

Description: This black, shiny-bodied native bee had a body length of 12 mm. Its wings were brownish, and the forewing length was nearly 8 mm. Mouthparts were partly reddish-yellow.

Approximate date of extinction: The species was collected once only, in 1938.

Probable cause of extinction: Since European settlement, the vegetation of Rottnest Island has changed dramatically from dense tall shrublands dominated by Rottnest Island tea-tree (*Melaleuca lanceolata*) and closed forests of Rottnest Island cypress (*Callitris preissii*) to low coastal scrub. This was caused by a combination of frequent fire and over-grazing by quokkas. The change in vegetation probably eliminated the food plants on which the native bee depended.

Kellerberrin snail *Bothriembryon praecelsus*

Description: The Kellerberrin snail was around 29 mm long and 20 mm wide. The opening of its shell was 17 mm long and 10 mm wide. The shell was very thin, inflated, and had a short, conical spire. The opening of the shell was a little longer than the spire. The shell was almost a uniform brown colour, with lighter growth lines.

Approximate date of extinction: The species has not been collected since the original specimens were taken from near Kellerberrin some time prior to 1939.

Probable cause of extinction: Land clearing.

Left The shell of the Kellerberrin snail.

Photo – Corey Whisson /WA Museum

Geraldton snail   *Bothriembryon whitleyi*

Description: The Geraldton snail was 16 mm long and 12.5 mm wide. It had a small, white shell, with an apex of two whorls, and an oval aperture.

Approximate date of extinction: This species has not been found again since the original series of shells were collected at Geraldton some time prior to 1939.

Probable cause of extinction: Land clearing and the introduction of exotic snails.

Albany snail   *Helicarion castanea*

Description: None available.

Approximate date of extinction: The first specimens of the Albany snail were located in 1826 near King George Sound. The species was also collected at Pemberton in 1955, and in the Geographe Bay area at about the same time, but has not been recorded since, despite searches by experts.

Probable cause of extinction: The introduction of the carnivorous snail *Oxychilus* sp. and other molluscs is thought to be the main cause of extinction. Land clearing may also have contributed to its demise.

Above Shells of the Albany snail.

Photo – Karen Edwards /WA Museum

Below The introduced predatory snail *Oxychilus* sp. has been implicated in the extinction of native snail species.

Photo – Jiri Lochman

Pemberton snail   *Occirhenea georgiana*

Other names: *Helix georgiana.*

Description: This carnivorous snail has a thin brown shell, which is wider than it is high. The small shell is 3.5 mm high and 7 mm wide.

Approximate date of extinction: The Pemberton snail was first collected at King George Sound in October 1826. It was not collected again until some were found in 1955, on the outskirts of Pemberton. It has not been seen since 1955, despite searching.

Probable cause of extinction: The introduction of the carnivorous snail *Oxychilus* sp. and other molluscs was the probable cause of its extinction. Land clearing may also have contributed.

Above Shells of the Pemberton snail.

Photo – Karen Edwards/WA Museum

Threatened species

To minimise repetition and promote understanding, some threatened invertebrates will be treated as species groups and others as individual species.

Threatened invertebrates of the North West Cape peninsula

Many subterranean animals are restricted to one or very few caves or groundwater features near Exmouth. Some exist within threatened ecological communities.

Two species of groundwater-dwelling (stygobitic) fish also occur on the peninsula—see Chapter 8.

Crustacean, no common name *Bunderia misophaga*

Crustacean, no common name *Danielopolina kornickeri*

Cape Range remipede *Lasionectes exleyi*

Crustacean, no common name *Stygocyclopia australis*

Crustacean, no common name *Speleophria bunderae*

Cape Range amphipod *Liagoceradocus branchialis*

Cape Range bristle worm *Prionospia thalanji*

These species all occur only in the Bundera Sinkhole on the western coastal plain of the North West Cape peninsula.

Together with other species, they form the 'Cape Range remipede community (Bundera Sinkhole)', a threatened ecological community that is discussed in Chapter 10. The ecological community, and the species restricted to it, are Critically Endangered.

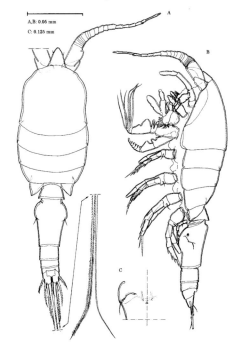

A,B: 0.05 mm
C: 0.125 mm

Above *Hyella humphreysi*, a pseudoscorpion closely related to the Camerons Cave pseudoscorpion *Hyella* sp.

Photo – Douglas Elford /WA Museum

Right The entrance to Camerons Cave.

Photo – Bill Humphreys /WA Museum

Camerons Cave millipede
Stygiochiropus peculiaris

Camerons Cave pseudoscorpion
Hyella sp.

These species have been found only in Camerons Cave, a small sinkhole south of Exmouth on the eastern coastal plain of North West Cape peninsula. Together with other species, they comprise 'Camerons Cave troglobitic community', a threatened ecological community that is discussed in Chapter 10. Both species are Critically Endangered, as is the ecological community.

Millipede, no common name   *Stygiochiropus isolatus*
Millipede, no common name   *Stygiochiropus sympatricus*

Description: These cave-dwelling millipedes are very small. *S. isolatus* is 7–8 mm long. The single specimen of *S. sympatricus* is 14 mm long.

Distribution and habitat: Each species is known only from a single, isolated cave on the North West Cape peninsula.

Biology and ecology: Nothing is known of their biology and ecology. It is thought that these millipedes are relictual forest forms that were isolated in caves, some time after the Miocene (23.8 to 5.3 million years ago), when the region became too dry to support forests.

Threats: The major threat to Cape Range and the North West Cape peninsula is limestone mining, as the limestone is of a high quality needed in mineral processing. Other threats include groundwater abstraction for the burgeoning tourism industry and damage to the caves and their environments.

Status: Vulnerable. The Department of Conservation and Land Management's North West Cape Karst Management Advisory Committee is the recovery team for these and other North West Cape threatened species and ecological communities.

Eastern Cape Range bamazomus   *Bamazomus subsolanus*

Western Cape Range bamazomus   *Bamazomus vespertinus*

Barrow Island draculoides   *Draculoides bramstokeri*

Northern Cape Range draculoides   *Draculoides brooksi*

Western Cape Range draculoides   *Draculoides julianneae*

Description: *Bamazomus* and *Draculoides* species are small spider-like schizomids. The Schizomida form a separate Order of arachnids and very few species occur in the southern hemisphere. These five species of cave-dwelling schizomids lack any eyes or eye spots.

Distribution and habitat: All five threatened schizomids inhabit caves on the North West Cape peninsula. The Barrow Island draculoides has been located in two caves on North West Cape peninsula and in caves on Barrow Island. The northern Cape Range draculoides and the western Cape Range bamazomus each occur in a single cave. The eastern Cape Range bamazomus was found in a limestone quarry.

Biology and ecology: Not known.

Threats: See above.

Status: The Barrow Island draculoides is Vulnerable; the other schizomids are Endangered. The Department of Conservation and Land Management's North West Cape Karst Management Advisory Committee is the recovery team for these and other North West Cape threatened species and ecological communities. The Barrow Island draculoides is discussed further under 'Threatened invertebrates of Barrow Island' (p. 151).

Lance-beaked cave shrimp   *Stygiocaris lancifera*

Description: This cave-dwelling atyid shrimp reaches up to 14 mm long. It is entirely colourless and transparent.

Distribution and habitat: The lance-beaked cave shrimp is endemic to the North West Cape peninsula and Barrow Island. On North West Cape, it occurs in several caves on the western coastal plain. The related species *Stygiocaris stylifera* occurs on the eastern coastal plain.

Biology and ecology: Not known.

Threats: Lowering of water tables and pollution threaten the lance-beaked cave shrimp. Some caves on the peninsula have already dried or been filled.

Status: Vulnerable. The Department of Conservation and Land Management's North West Cape Karst Management Advisory Committee is the recovery team for these and other North West Cape threatened species and ecological communities.

Threatened spiders of the Nullarbor

Mullamullang cave spider   *Tartarus mullamullangensis*

Murdoch Sink cave spider   *Tartarus murdochensis*

Nurina cave spider   *Tartarus nurinensis*

Thampanna cave spider   *Tartarus thampannensis*

Nullarbor cave trapdoor spider   *Troglodiplura lowryi*

Description: A number of troglobitic spiders inhabit caves of the Nullarbor Plain.

Distribution and habitat: These spiders live in caves in the Nullarbor region. The four species of *Tartarus* are each restricted to a single cave, indicated by their common name. The

Below *Tartarus* species (top) and the Nullarbor cave trapdoor spider.

Photos – Paul Devine

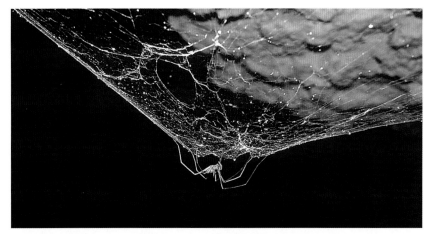

Mullamullang cave spider has been found up to 4 km inside the cave entrance. The Nullarbor cave trapdoor spider is found in Roach's Rest Cave and Old Homestead Cave, both north-east of Madura, and in a cave near the head of the Bight in South Australia. It is the only known cave-dwelling mygalomorph (member of a group of primitive spiders that include the trapdoor spiders) in Australia.

Biology and ecology: The *Tartarus* spiders build broad funnel-like webs, within which the spider sits head down and motionless on the rock wall of the cave. It is thought that the web forms a snare adapted to catch walking prey, probably arthropods such as beetles, cockroaches, centipedes and isopods that also occur in the caves. The Nullarbor cave trapdoor spider apparently makes no web, but hunts prey while moving around the cave.

Threats: All species are listed primarily because they have very restricted distributions and are uncommon within their habitat. They are threatened because the caves are visited by people and because the Nullarbor Plain has suffered significant degradation—due to removal of vegetation by rabbits and stock, as well as weed invasion—that has resulted in a greater flow of silt-loaded water into the caves than was formerly the case.

Status: All these spiders are listed as Vulnerable.

Threatened invertebrates of Barrow Island

Barrow Island is composed of limestone, with many caves and underground cavities. It supports a rich subterranean fauna and new species are still being discovered. As well as the listed species, there are many other poorly-sampled and poorly-known species of stygofauna, including copepods and other crustaceans. At least 25 species of stygofauna are found on the island, and probably many more. Most of them live in a narrow lens of fresh water that lies over deep saline waters beneath the island. The terrestrial animals—the millipede and the *Draculoides*—live in caves.

The Barrow Island draculoides (together with the blind gudgeon) also occurs on the North West Cape peninsula—Barrow Island was joined to North West Cape prior to about 8000 years ago. Other species are thought to be endemic to Barrow Island, but recent research by Department of Conservation and Land Management scientists, as part of the Pilbara Region Biological Survey, suggests that some of the stygofauna may occur on the adjacent mainland.

TROGLOFAUNA

Barrow Island millipede   *Speleostrophus nesiotes*

Description: This troglobitic millipede is about 28 mm long.

Distribution and habitat: The Barrow Island millipede has been found only from an area of a few square metres within a single cave on Barrow Island. A recent visit to the cave failed to locate any millipedes. This millipede belongs in a group that, in Australia, occurs in rainforests in Queensland and New South Wales, although its closest relatives appear to be from India.

Biology and ecology: Millipedes live in soil and humus.

Threats: Visitors to the cave could easily trample and damage the habitat of this species. ChevronTexaco, which manages the oilfield on Barrow Island, has made the cave out of bounds to its staff and contractors.

Status: Endangered. Research is urgently needed to clarify whether this species occurs elsewhere on Barrow Island and whether it still occurs within the known cave.

Barrow Island draculoides   *Draculoides bramstokeri*

Description: This small, light yellow to brown troglobitic schizomid has a body length up to 5 mm.

Distribution and habitat: The Barrow Island draculoides is known from six caves on Barrow Island and two on North West Cape peninsula.

Biology and ecology: Not known.

Threats: Damage to or pollution of caves are potential threats.

Status: Vulnerable. The North West Cape Karst Management Advisory Committee coordinates the conservation of threatened animals on the North West Cape peninsula.

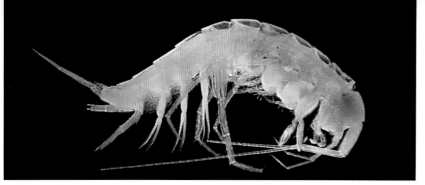

STYGOFAUNA

Barrow Island bogidomma amphipod   *Bogidomma australis*

Cape Range amphipod   *Liagoceradocus branchialis*

Barrow Island liagoceradocus amphipod   *Liagoceradocus subthalassicus*

Fragile amphipod   *Nedsia fragilis*

Humphrey's amphipod   *Nedsia humphreysi*

Hurlbert's amphipod   *Nedsia hurlberti*

Amphipod, no common name   *Nedsia macrosculptilis*

Amphipod, no common name   *Nedsia sculptilis*

Amphipod, no common name   *Nedsia straskraba*

Amphipod, no common name   *Nedsia urifimbriata*

Description: These species of subterranean amphipod crustaceans vary in length from 2.5–6 mm and are white or translucent.

Distribution and habitat: The stygobitic amphipods inhabit a narrow lens of fresh water that sits above salt groundwater under the island.

Biology and ecology: Bill Humphreys of the WA Museum has speculated that the biological diversity seen in the Barrow Island stygofauna may, in part, be supported by, or even dependent upon, the petroleum deposit below the island. The Barrow Fault, which reaches the surface towards the southern end of the island, releases hydrogen sulphide in vents at the surface and into the groundwater, where sulphur bacteria are likely to carry out chemoautotrophic energy production. (A chemoautotrophic organism obtains its nourishment through the oxidation of inorganic chemical compounds, as opposed to photosynthesis.)

Threats: The production of oil since the 1960s has resulted in considerable pollution of the groundwater through disposal of 'produced water' (saline water that is mixed with the underground oil, separated from the oil on the surface and then reinjected, but with traces of oil remaining), and accidental oil, chemical and salt water spills. In recent years, most 'produced water' has been injected deeper underground. What effect this pollution has had, if any, is not known.

Status: All listed Barrow Island stygofauna are Vulnerable.

Notes: One stygobitic fish species, the blind gudgeon (*Milyeringa veritas*), also occurs on Barrow Island (see p. 140). What is thought to be the world's only species of troglobitic reptile, *Ramphotyphlops longissimus*, is known from a single specimen from Barrow Island. It is included in the Department of Conservation and Land Management's List of Priority Fauna, as there have been no comprehensive searches targeting it.

Threatened land snails of the Ningbing Range and Jeremiah Hills

The following species of snail, all in the family Camaenidae, are listed as threatened. None have common names.

Cristilabrum bubulum

Cristilabrum buryillum

Cristilabrum grossum

Cristilabrum isolatum

Cristilabrum monodon

Cristilabrum primum

Cristilabrum rectum

Cristilabrum simplex

Cristilabrum solitudum

Cristilabrum spectaculum

Ningbingia australis australis

Ningbingia australis elongata

Ningbingia bulla

Ningbingia dentiens

Ningbingia laurina

Ningbingia octava

Ningbingia res

Ordtrachia elegans

Turgenitubulus christenseni

Turgenitubulus costus

Turgenitubulus depressus

Turgenitubulus foramenus

Turgenitubulus opiranus

Turgenitubulus pagodula

Turgenitubulus tanmurrana

Description: The Camaenidae is a diverse group of terrestrial hermaphroditic pulmonate (having lungs or lung-like organs) molluscs found in the tropics of both the eastern and western hemispheres. Eastern Asia and the Australasian region have the highest diversity of camaenid species. Many Kimberley species have shells more than 25 mm across, with some being up to 35 mm in diameter.

Distribution and habitat: Each of these land snails is known from only one or two locations in the Ningbing Range or Jeremiah Hills, north of Kununurra. All occur in small, isolated 'vine thickets' (remnant rainforest patches) in valleys in the limestone range, which provide shade, cover and cooler, moister conditions than surrounding areas. The genera *Ningbingia*, *Turgenitubulus* and *Cristilabrum* are restricted to the Ningbing Range and Jeremiah Hills. This is possibly the greatest concentration of short range restricted endemic species anywhere in the world, with more than 28 species packed into 52 km of limestone hills. The median linear range of a species is 1.65 km, and the median area range is 0.825 km^2.

Biology and ecology: These snails feed on algal and fungal films on rocks, wood and so on. Snails reach half their adult size in the wet season of their birth, then reach adult shell size and become mature males at the end of their second wet season. They function as males by the beginning of their third wet season. The female genitals mature near the end of this season and the snails function as both males and females in their fourth and subsequent wet seasons. They can live for more than eight years.

Above Shells of *Turgenitubulus costus*.

Photo – Karen Edwards/ WA Museum

Threats: The threats to all species are the same—the vine thickets are degrading and disappearing due to frequent fire and cattle grazing.

Status: All but six of these land snail species are Critically Endangered. *Cristilabrum bubulum, C. isolatum, C. spectaculum* and *Turgenitubulus christenseni* are Endangered and *Turgenitubulus pagodula* and *Ordtrachia elegans* are Vulnerable.

These snails were first listed in 2003. The Department of Conservation and Land Management will be preparing a single Interim Recovery Plan covering all species, and will work with the pastoral station lessee to minimise further habitat damage and attempt to implement management that will allow the vine thickets to regenerate and expand.

Right Shells of *Turgenitubulus pagodula* (top) and *Ordtrachia elegans* (below).

Photo – Karen Edwards /WA Museum

Threatened land snails of the Napier Range

Snail, family Camaenidae, no common name *Mouldingia occidentalis*

Snail, family Camaenidae, no common name *Westraltrachia alterna*

Snail, family Camaenidae, no common name *Westraltrachia inopinata*

Snail, family Camaenidae, no common name *Westraltrachia turbinata*

Description: The Camaenidae is a diverse group of terrestrial hermaphroditic pulmonate (having lungs or lung-like organs) molluscs found in the tropics. Eastern Asia and the Australasian region have the highest diversity of Camaenid species. Many Kimberley species have shells more than 25 mm across, with some being up to 35 mm in diameter.

Distribution and habitat: *Mouldingia occidentalis* and *Westraltrachia alterna* are confined to McSherry Gap, while *Westraltrachia inopinata* and *W. turbinata* are confined to Yammera Gap. Like those of the Ningbing Range and Jeremiah Hills (see p. 153), these land snails inhabit small areas of vine thicket.

Biology and ecology: See Threatened Land Snails of the Ningbing Range and Jeremiah Hills.

Threats: Frequent fire and grazing, leading to habitat degradation, are serious threats.

Status: *Mouldingia occidentalis* is Critically Endangered; and the Department of Conservation and Land Management will be preparing an Interim Recovery Plan aimed at conserving the species. The three species of *Westraltrachia* are Vulnerable.

Left Shell of *Mouldingia occidentalis.*

Photo – Karen Edwards /WA Museum

Individual species

INSECTS

Short-tongued bee   *Leioproctus douglasiellus*

Description: This small, short-tongued native bee has a body length of about 8 mm. Its body is black. The wings are greyish, with dark brown veins, and the mouthparts are reddish-yellow.

Distribution and habitat: The short-tongued bee has been collected at Pearce and Forrestdale Lake on the Swan Coastal Plain near Perth.

Biology and ecology: Its life history is unknown, but it seems to be dependent on the flowers of narrow-leaved goodenia (*Goodenia filiformis*), a rare plant that occurs on the Swan Coastal Plain and on the south coast between Cape Riche west to near Manjimup. It has also been recorded feeding on *Anthotium junciforme*, which occurs from Perth south to Augusta.

Threats: Habitat clearing.

Status: Endangered. Further searches and study are required before the conservation requirements of this species can be determined.

Native bee, no common name   *Neopasiphae simplicior*

Description: This small native bee has a body length of about 7 mm and a wing length of 5 mm. It is creamy yellow and brown in colour.

Distribution and habitat: The species is found only near Perth. It has been collected at Cannington and near Forrestdale Lake around the Forrestdale golf course.

Biology and ecology: Its life history is unknown. Females have been collected on flowers of narrow-leaved goodenia (*Goodenia filiformis*) and slender lobelia (*Lobelia tenuior*). Males have been collected on Preiss's angianthus (*Angianthus preissianus*).

Threats: Land clearing.

Status: Endangered. Further searches and study are required before conservation requirements can be ascertained.

Graceful sunmoth *Synemon gratiosa*

Description: Sunmoths are medium-sized, day-time-flying moths with brightly-coloured hindwings and clubbed antennae. The graceful sunmoth has orange hindwings.

Distribution and habitat: The graceful sunmoth is known from just 13 locations on the Swan Coastal Plain near Perth, from Neerabup to Mandurah. However, in the last decade it has been recorded only from Neerabup, Whiteman Park and the Koondoola bushland.

Biology and ecology: Sunmoth larvae feed on the underground parts of sedges and grasses such as *Lepidosperma* and *Lomandra* species. The food plants of the graceful sunmoth larvae are as yet unknown, but species of these two plant genera are reasonably common near Perth. Adults are thought to feed on flowers.

Threats: Land clearing for urban development.

Status: Endangered. The Department of Conservation and Land Management has supported a grant application by the Western Australian Insect Study Society to search for and study populations of this species.

Above In the last decade, the graceful sunmoth has been found only at three localities near Perth.

Photo – Terry Houston /WA Museum

MILLIPEDES

Western Australian pill millipede *Cynotelopus notabilis*

Description: Pill millipedes are distinguished from other millipedes by their extremely stout body that is capable of rolling into a tight ball. The WA pill millipede is the only species in this group known from the State, and has a ball width of 6.3–7.6 mm. Its general colouration is greenish-black.

Distribution and habitat: The WA pill millipede has a very restricted range near the south coast, from Tinglewood east to Torbay Hill. It inhabits deep litter or moist areas under rotting logs and rocks.

Biology and ecology: Not known. Millipedes eat rotting leaves and wood, fungi and algae.

Threats: Some habitat has been cleared for agriculture. All remaining known localities are in State forest or national parks. As the species occurs in deep leaf litter, in long-unburnt areas and similarly sheltered sites, inappropriate fire regimes are likely to be a threat. Pill millipedes, like many other millipedes, lack any significant dispersive stage and are vulnerable to local extinction events. The species seems to have once been much more common in karri forest than it is now. Research into its ecology is needed.

Status: Endangered. WA pill millipedes occur within Walpole-Nornalup, William Bay and West Cape Howe national parks.

Above Western Australian pill millipede.

Photo – Jane McRae/CALM

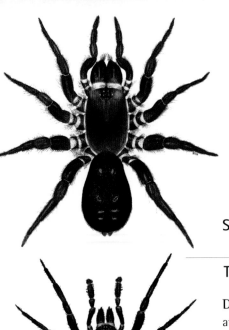

Above The female (top) and male tree-stem trapdoor spiders.

Illustrations – Brad Durrant

SPIDERS AND SCORPIONS

Tree-stem trapdoor spider   *Aganippe castellum*

Description: Tree-stem trapdoor spiders are medium-sized spiders with a carapace length of about 12 mm. They dig their burrows against the stems of shrubs or trees, usually broombush (*Melaleuca uncinata*) or sheoaks such as *Allocasuarina acutivalvis*, but also on various myrtaceous shrubs and occasionally eucalypts. The silk tube lining is extended above the ground and is attached to the tree or shrub, always with the rim facing up the stem and the door opening away from the stem. A fan of twig-lines is attached to the rim, and the free ends of these twigs hang down to the ground. This nest structure is unique among trapdoor spiders (see p. 142).

Distribution and habitat: Tree-stem trapdoor spiders occur throughout the central, eastern and northern Wheatbelt in wodgil (acacia shrubland, especially *A. neurophylla*) habitats. The soil is generally a sandy loam and the habitat may be subject to sheet flooding in winter.

Biology and ecology: Females spend their entire lives—20 years or more—in the same burrow, while males leave their burrows only to search for mates following autumn thunderstorms.

Threats: Clearing for agriculture has destroyed most of their habitat. The few, small, remaining populations could easily be destroyed through road building and gravel extraction, or by hot fires. Recruitment is slow, and dispersing juveniles cannot move between vegetation remnants.

Status: Endangered. Small populations still occur in remnant vegetation, including some small nature reserves.

Western archaeid spider   *Austrarchaea mainae*

Description: The western archaeid spider is very small (about 3.7 mm long), but has very long fangs.

Distribution and habitat: The species is known only from Torndirrup National Park, near Albany, where it lives in deep litter in thickets of dwarf peppermint (*Agonis flexuosa*). Other members of this genus live in rainforests.

Biology and ecology: Not known.

Threats: Western archaeid spiders are probably vulnerable to fire, as they were first collected in a site not burnt for 17 years.

Status: Vulnerable. The recently-formed South Coast Threatened Invertebrate Group is working to conserve the western archaeid spider and other threatened invertebrates from the region.

Shield-backed trapdoor spider
Idiosoma nigrum

Description: The shield-backed trapdoor spider has a carapace that is 10 mm long and 7 mm wide. It is dark greyish-brown, sometimes with a reddish tinge. The burrow is lined with white silk and a fan of twigs is always attached to the rim. The twigs are usually arranged in a 'moustache' fashion, that is, there are two bundles with a gap in between. The door is made of litter fragments bound with silk. The entrance is a vertical cup. If it is disturbed, the spider may block the narrow burrow entrance at the base of the 'cup' with its hard abdomen.

Distribution and habitat: The spider occurs throughout the mid-west, south to Toodyay, Northam and Beverley, extending northwards to Nanga. Nests are usually located in litter within acacia woodland or shrubland, particularly in jam (*Acacia acuminata*) on granitic soils, but also in eucalypt woodlands on heavy soils. The burrow is cup-shaped at the top and then narrows for some distance, before widening again towards the base. Burrows are lined with silk, and the lid is wafer like, easily bent and has leaves, twigs and bark attached to its upper surface.

Biology and ecology: Ants appear to be a major food item, as ant remains are found in the bottom of burrows.

Threats: Clearing for agriculture has destroyed most of this species' habitat.

Status: Vulnerable. Some populations occur on nature reserves.

Above Shield-backed trapdoor spider.

Photo – Barbara York Main

Below The burrow entrance of the shield-backed trapdoor spider.

Yorkrakine trapdoor spider   *Kwonkan eboracum*

Description: Adult Yorkrakine trapdoor spiders are relatively small, with a carapace of 4.3 mm (males) and 7 mm (females). The long, thin legs and the body are sparsely hairy and shiny yellow, with prominent brown bands on the abdomen.

Distribution and habitat: The Yorkrakine trapdoor spider was first found on 'Eboracum', a property near Tammin, and was named after it by spider expert Barbara York Main. It was later found at nearby Yorkrakine Rock. Despite further surveys, including a detailed search in 1999 funded by the Natural Heritage Trust, no more were found. In 2000, however, a very small population was located in a road verge adjacent to the type locality (now cleared) at Eboracum. The type locality was a kwongan (heath) shrubland on yellow sand adjacent to open salmon gum woodland. The road verge has a few scattered eucalypts with an open understorey of hummock-like shrubs of *Acacia, Allocasuarina* and native grasses. The original range is unknown, but the species was probably restricted to a relatively small area of the Wheatbelt.

Biology and ecology: The Yorkrakine trapdoor spider lives in a shallow (less than 20 cm deep) vertical burrow lined with silk.

Threats: Clearing has destroyed most of the spider's habitat. Being in a road verge, the only known, small population needs protection from damage by machinery and fire.

Status: Critically Endangered. The Department of Conservation and Land Management and Tammin Shire are protecting the road verge. Yorkrakine Rock is a nature reserve, but the spider has not been found there for some time.

Above The burrow of a Stirling Range moggridgea (*Moggridgea* sp.).

Above right Stirling Range moggridgea spider.

Photos – Jiri Lochman/CALM

Stirling Range moggridgea spider   *Moggridgea* sp.

Description: These small spiders (to a total length of about 8 mm) are shiny dark brown to almost black, and have a slightly humped abdomen. They live in short burrows (up to 6 cm long) with lids that sometimes incorporate moss, algae or liverwort growth as well as soil particles. Burrows are usually in heavy soils in creek banks, and tend to occur in aggregations within small areas of up to 3 m². Another species of *Moggridgea* occurs in the Porongurup Range. It is unclear whether the Stirling Range moggridgea represents a single species or whether there are several species, each restricted to particular peaks.

Distribution and habitat: All known populations are small and fragmented, occurring on shaded, south-facing slopes and valleys in the Stirling Range.

Other *Moggridgea* species occur in southern Africa and South Australia and they are true Gondwanan relics.

Biology and ecology: Nothing is known about the biology and ecology of the species, but it is probably similar to other trapdoor spiders.

Threats: The burrows are too shallow for their occupants to withstand the heat generated at ground level by bushfires, and fire is the main threat to the spider's survival.

Status: Endangered. The Department of Conservation and Land Management's recently-formed South Coast Threatened Invertebrate Group is working to conserve this and other threatened invertebrates of the region.

Above Adults of the tingle trapdoor spider are only about 1 cm in length.

Photo – Barbara York Main

Tingle trapdoor spider   *Moggridgea tingle*

Description: The female has a carapace length of 2.6–3.1 mm.

Distribution and habitat: The tingle trapdoor spider is known from only four locations, all in tingle (*Eucalyptus jacksonii*) forest in Walpole-Nornalup National Park.

Biology and ecology: Nests have been found in soil, or in the fibrous bark of tingle trees, and are constructed as silk tubes bound with particles. The tubes are up to 2 cm long, or just over, and are located outside the soil or bark. The opening usually faces upslope, with the hinge line away from the surface on which it is built. Nests are in shaded, damp sites.

Threats: Frequent fire and habitat destruction through road building are the main threats.

Status: Endangered. The Department of Conservation and Land Management has developed a fire response plan to ensure that the spider's habitat is not damaged during firefighting operations. The known locations are plotted on management maps to prevent damage during park operations.

Minnivale trapdoor spider *Teyl* sp.

Description: Minnivale trapdoor spiders are small to medium-sized, with a total body length—excluding legs and pedipalps (appendages near the mouth of a spider or other arachnid that are modified for various reproductive, predatory, or sensory functions)—of 11 mm in males and 14 mm in females. The legs are long and thin, and the body segments are sparsely hairy. The general colour of the body and legs is a dull, dusty tan. Most species of *Teyl* have simple, open-holed burrows, but several species, including this one, build trapdoors—an adaptation against sheet flooding. The vertical burrow is closed at the surface with a door, and has a side shaft that is also closed by a door. Both doors are made of compacted soil. They are circular and flat on the 'outside', with the underside rounded and silk-covered. The surface door fits snugly into the vertical burrow opening, and, when open, the hinged door lies flat ('upside down') on the ground. The side shaft door, when open, hangs downward from the horizontal hinge, into the main shaft of the burrow. The burrow is deep (the first excavated burrow was 33 cm deep) with a close-fitting plug-shaped lid, which is extremely difficult to see when it is closed.

Distribution and habitat: The Minnivale trapdoor spider is known from two localities in the northern and central Wheatbelt of WA: on Mellanbye Station north-east of Gutha and near Minnivale. Old, disused burrows and doors, tentatively identified as having been constructed by this species, have been found at North Bungulla Nature Reserve and at Gutha. The former distribution of the species is thought to have been over a narrow band, roughly between Minnivale and Mellanbye in 'perched'

swamps on high terrain. It probably did not extend far westward, as another rare *Teyl* species occurs in wandoo country, while several undescribed species occur further to the east in salt lake country of the eastern Wheatbelt and braided creeks of the eastern Goldfields (from Wydgee, near Mount Magnet, to Scadden, north of Esperance).

Biology and ecology: The behaviour of *Teyl* species is seasonal, as they are active in winter. One nest was found with young near Minnivale. The only known male was collected as a penultimate instar, and was kept in a flower pot until it surfaced as an adult in September.

Threats: Land clearing has been the main threat to this species, with most of its presumed geographic range having been cleared for agricultural use.

Status: Critically Endangered. The Department of Conservation and Land Management prepared an Interim Recovery Plan in 1999. At that time, a single occupied burrow was known in an area of remnant vegetation near Minnivale. The plan's main actions were to protect this burrow and search for additional populations. Since then, the spider in the burrow has either died or, if it was a male, left the burrow during the mating season. In 1999 and 2001, extensive searches carried out by spider experts Barbara York Main and Mark Harvey, with the assistance of Julianne Waldock, did not uncover any new populations. Thus, the species may be extinct.

Above Male (top) and female Minnivale trapdoor spiders vary in size and colour.

Illustrations – Brad Durrant

SNAILS

Stirling Range rhytidid snail Undescribed snail in the family Rhytididae

Description: None available.

Distribution and habitat: This snail is known only from two localities in the Stirling Range: the Cascades to the north of Bluff Knoll and south of Ellen Peak. It inhabits shaded, humid gullies with running streams, dense tree cover, deep leaf litter and ferns, and shelters beneath rocks and logs.

Biology and ecology: No information is available on its biology. The snail is probably carnivorous, predating other snails such as *Bothriembryon glauerti*, with which it also occurs.

Threats: Hot wildfires threaten this snail and other species restricted to similar habitats, such as the trapdoor spider *Neohomogona* sp. and the pseudoscorpion *Pseudotyrannochthonius* sp. All these species are thought to have survived from early Tertiary rainforests. In recent years, there have been several hot fires in Stirling Range National Park, and the Cascades area was badly affected in 1996.

Status: Endangered. The Department of Conservation and Land Management is attempting to protect the habitat of this species from fire. The species is highly threatened, and may be reclassified as Critically Endangered.

Above The shell of the Cape Leeuwin freshwater snail.

Photo – Karen Edwards /WA Museum

Cape Leeuwin freshwater snail *Austroassiminea letha*

Description: The always conical shell of this species is nonetheless very variable in shape, from fairly squat to more elongate. Shells of males are often smaller and squatter than those of females. The shell is 3.45–5.39 mm (mean 4.50 mm) high.

Distribution and habitat: The Cape Leeuwin freshwater snail inhabits seepage films or splash zones alongside small freshwater streams draining from limestone near the coast north of Augusta from Ellen Brook south to Turner Brook in the Leeuwin-Naturaliste National Park. Aestivating individuals have been found on logs, leaves and rocks immediately adjacent to areas where they are active. Fossil collections indicate that the snail once had a more extensive distribution within the Leeuwin-Naturaliste Ridge.

Biology and ecology: Nothing known.

Threats: As the snail is currently known from only three localities, any activity that affected the amount of—or quality of—water in the streams could detrimentally impact on this highly restricted species. One population is under threat from the introduced predatory snail *Oxychilus* sp.

Status: Vulnerable.

CRUSTACEANS

Pannikin Plain cave isopod   *Abebaioscia troglodytes*

Description: A troglobitic isopod crustacean.

Distribution and habitat: This isopod is known only from Pannikin Plain Cave in the Nuytsland Nature Reserve, west of Madura, near the southern edge of the Nullarbor Plain.

Biology and ecology: Not studied.

Threats: Threats have not been studied, but, as the species is known only from a single cave, it could be threatened by damage to that cave.

Status: Vulnerable.

Crystal Cave crangonyctoid   *Hurleya* sp.

Description: None available.

Distribution and habitat: The Crystal Cave crangonyctoid occurs only in Crystal Cave in Yanchep National Park.

Biology and ecology: This troglobitic species lives in pools of water, now being artificially maintained, in Crystal Cave.

Threats: Pools in the caves in Yanchep National Park have been greatly reduced, or have dried, because of falling groundwater levels (see p. 170).

Status: Critically Endangered. Action being taken to try to save the Crystal Cave crangonyctoid, and other animals that depend on water in the Yanchep caves, is described in Chapter 10.

Above The Crystal Cave crangonyctoid found within Yanchep National Park is Critically Endangered.

Photo – Edyta Jasinska

References

Blyth, J. and Abbott, I. (1991). Spineless wonders. Are invertebrates second class citizens? *LANDSCOPE* 6(3), 28–33.

Burbidge, A.A., Harvey, M. and Main, B.Y. (1999). Minnivale Trapdoor Spider Interim Recovery Plan 1998–2000. Interim Recovery Plan No. 19. Department of Conservation and Land Management, Wanneroo.

Humphreys, W.F. (2001). The subterranean fauna of Barrow Island, northwestern Australia, and its environment. *International Journal of Subterranean Biology* 28, 107–127.

Main, B.Y., Harvey, M.S. and Waldock, J.M. (2002). The distribution of the Western Australian pill millipede *Cynotelopus notabilis* Jeekel (Sphaerotheriidae). *Records of the Western Australian Museum* 20, 383–385.

Solem, A. (1988). Maximum in the minimum: Biogeography of land snails from the Ningbing Ranges and Jeremiah Hills, northeast Kimberley, Western Australia. *Journal of the Malacological Society of Australia* 9, 59–113.

Yen, A. and Butcher, R. (1997). *An overview of the conservation of non-marine invertebrates in Australia.* Environment Australia, Canberra.

Threatened ecological communities

Biodiversity has three components—genetic diversity, species diversity and ecosystem diversity. To be effective, biodiversity conservation programs must address all three elements.

Many conservation biologists have argued that the major thrust of biodiversity conservation should be aimed at ecosystems, rather than species. However, while the declaration of reserves (or protected areas) has been a major part of nature conservation for a long time, and is aimed partly at protecting ecosystems, only recently have programs started to look at the conservation of individual threatened ecosystems. Ecological communities are the biological part of ecosystems—an ecosystem also has non-biological components, such as soil type, hydrological features, climate, topography and so on.

While there is currently no legislation in Western Australia covering the conservation of ecological communities, the Commonwealth first enacted such legislation in 1992. In 1994, the Department of Conservation and Land Management started to develop procedures for evaluating the conservation status of ecological communities in WA. With the aid of a Threatened Ecological Communities Scientific Committee, and after wide consultation with scientists, and business and community groups, definitions were developed for 'ecological community' and for categories of threat similar to those developed by the IUCN for species (see p. 167).

The definition of 'ecological community' used in WA is 'a naturally occurring biological assemblage that occurs in a particular type of habitat'.

It was adapted from the Victorian *Flora and Fauna Guarantee Act 1988*, and is close to the definition in the Commonwealth *Environment Protection and Biodiversity Conservation Act 1999*.

The next step was to begin evaluating particular WA communities to see if they met criteria to be regarded as threatened. Lists of possible threatened ecological communities were compiled from suggestions by scientists and naturalists, and from published scientific papers. These were then reviewed by Departmental staff and submitted to the Scientific Committee. The program has been ongoing since, with many more ecological communities nominated.

Before the Scientific Committee recommends a listing, it needs to be convinced that:

- the community is described so that all variants of it clearly fit the description, and it is distinct from all other assemblages, locally or anywhere else;

- sufficient information about the distribution of the community is available, and that it has been searched for adequately; and

- sufficient information is available to decide whether the community falls into one of the categories of threat (if not, the community would be classified as data deficient).

In April 2003, the Department's database contained 69 threatened ecological communities endorsed as such by the Minister for the Environment.

Above Bundera Sinkhole is the only deep, tidally-influenced karst system known in Australia.

Photo – Peter Kendrick/CALM

Opposite The Vulnerable threatened ecological community at Roebuck Bay is unique in the world. Of the dozen or so areas in the world containing huge intertidal flats rich in shorebirds, it is the most species-rich in terms of both birds and the invertebrates on which they feed.

Photo – Jan van de Kam

Above A troglobitic harvestman (*Glennhuntia glennhunti*) that occurs only in Camerons Cave.

Below A troglobitic micro-whipscorpion (*Draculoides* sp.).

Photos – Douglas Elford /WA Museum

Of these, 21 were Critically Endangered, 17 were Endangered, 28 were Vulnerable and three were Presumed Destroyed. Comparisons between States, and between WA and the whole of Australia, are not of any value at this early stage of development of the listing of threatened ecological communities.

Though most threatened ecological communities comprise plants, animals and other organisms, most have been described in terms of their vegetation. Twelve, however, are based primarily on animals, and these are discussed below. They are mostly stygofaunal or mound spring communities.

Definitions and criteria for Presumed Totally Destroyed, Critically Endangered, Endangered and Vulnerable Ecological Communities

Presumed Totally Destroyed (PD)

An ecological community which has been adequately searched for but for which no representative occurrences have been located. The community has been found to be totally destroyed, or so extensively modified throughout its range that no occurrence of it is likely to recover its species composition and/or structure in the foreseeable future.

An ecological community will be listed as Presumed Totally Destroyed if there are no recent records of the community **and either**:

A) records within the last 50 years have not been confirmed despite thorough searches of known or likely habitats; or

B) all occurrences recorded within the last 50 years have since been destroyed.

Critically Endangered (CR)

An ecological community that has been adequately surveyed and found to have been subject to a major contraction in area and/or that was originally of limited distribution, and is facing severe modification or destruction throughout its range in the immediate future, or is already severely degraded throughout its range but capable of being substantially restored or rehabilitated.

An ecological community will be listed as Critically Endangered when it has been adequately surveyed, and is found to be facing an extremely high risk of total destruction in the immediate future. This will be determined on the basis of the best available information, by it meeting **any one or more** of the following criteria (A, B or C).

A) The estimated geographic range, and/or total area occupied, and/or number of discrete occurrences since European settlement have been reduced by at least 90% **and either or both** of the following apply:

i) geographic range, and/or total area occupied and/or number of discrete occurrences are continuing to decline, such that total destruction of the community is imminent (within approximately 10 years);

ii) modification throughout its range is continuing such that in the immediate future (within approximately 10 years) the community is unlikely to be capable of being substantially rehabilitated.

B) Current distribution is limited, **and one or more** of the following apply:

i) geographic range and/or number of discrete occurrences, and/or area occupied is highly restricted, and the community is currently subject to known threatening processes which are likely to result in total destruction throughout its range in the immediate future (within approximately 10 years);

ii) there are very few occurrences, each of which is small and/or isolated and extremely vulnerable to known threatening processes;

iii) there may be many occurrences, but total area is very small, and each occurrence is small and/or isolated and extremely vulnerable to known threatening processes.

C) The ecological community exists only as highly modified occurrences that may be capable of being rehabilitated if such work begins in the immediate future (within approximately 10 years).

Below Tuart forest above Cabaret Cave, Yanchep National Park (see p. 170). Photo – Michael James/CALM

Endangered (EN)

An ecological community that has been adequately surveyed and found to have been subject to a major contraction in area and/or was originally of limited distribution and is in danger of significant modification throughout its range, or severe modification or destruction over most of its range, in the near future.

An ecological community will be listed as Endangered when it has been adequately surveyed and is not Critically Endangered, but is facing a very high risk of total destruction in the near future. This will be determined on the basis of the best available information by it meeting **any one or more** of the following criteria (A, B or C).

A) The geographic range, and/or total area occupied, and/or number of discrete occurrences have been reduced by at least 70% since European settlement and **either or both** of the following apply:

   i) the estimated geographic range, and/or total area occupied and/or number of discrete occurrences are continuing to decline such that total destruction of the community is likely in the short term (within approximately 20 years);

   ii) modification throughout its range is continuing such that in the short term (within approximately 20 years) the community is unlikely to be capable of being substantially restored or rehabilitated.

B) Current distribution is limited, and **one or more** of i, ii or iii apply:

   i) geographic range and/or number of discrete occurrences, and/or area occupied is highly restricted and the community is currently subject to known threatening processes which are likely to result in total destruction throughout its range in the short term (within approximately 20 years);

   ii) there are few occurrences, each of which is small and/or isolated, and all or most occurrences are very vulnerable to known threatening processes;

   iii) there may be many occurrences, but total area is small and all or most occurrences are small and/or isolated and very vulnerable to known threatening processes.

C) The ecological community exists only as very modified occurrences that may be capable of being substantially restored or rehabilitated if such work begins in the short term (within approximately 20 years).

Vulnerable (VU)

An ecological community that has been adequately surveyed, and is found to be declining and/or has declined in distribution and/or condition, and whose ultimate security has not yet been assured and/or a community that is still widespread but is believed likely to move into a category of higher threat in the near future if threatening processes continue or begin operating throughout its range.

An ecological community will be listed as Vulnerable when it has been adequately surveyed and is not Critically Endangered or Endangered, but is facing a high risk of total destruction or significant modification in the medium to long term future. This will be determined on the basis of the best available information by it meeting **any one or more** of the following criteria.

A) The ecological community exists largely as modified occurrences that are likely to be capable of being substantially restored or rehabilitated.

B) The ecological community may already be modified and would be vulnerable to threatening processes, is restricted in area and/or range and/or is only found at a few locations.

C) The ecological community may be still widespread but is believed likely to move into a category of higher threat in the medium to long term because of existing or impending threatening processes.

Below Roebuck Bay. Photo – Jan van de Kam

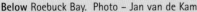

Threatened ecological communities

Communities of tumulus springs
(Organic mound springs of the Swan Coastal Plain)

Description: Tumulus (mound) springs occur on the eastern Swan Coastal Plain near Perth, where the clays of the Guildford formation adjacent to the Darling Scarp meet the Bassendean Sands, at the edge of the Gnangara Mound superficial aquifer. Areas of groundwater discharge—at the junction of the permeable sands and the impermeable clays—form springs, bogs and swamps. In the case of the tumulus springs, continuous vegetation growth leads to the formation of peat around the permanent water supply. Water continues to penetrate the increasingly elevated peat layers, due to the pressure created by local and regional hydrological forces. Where water finds a preferred pathway or conduit through the soil, water movement is much faster than normal groundwater flow. Such conduits, or pipes, may carry sand and silt to the surface, where it is deposited as a 'collar' of increasing height, thus enhancing the formation of mounds.

Distribution: These communities were once common within a narrow range, but there are now only three remaining vegetated occurrences: near Ellenbrook, near Bullsbrook and near Muchea.

Biological characteristics: The tumulus springs are permanently moist, and some are also associated with permanent pools and surface water. Many of the invertebrate animals and the plant species found in these communities are adapted to this permanent moisture, and the areas probably act as refuges from climate change (increasing aridity) for certain species. Some of the invertebrate species would not survive if the peat mounds were to dry out.

Threats: Clearing for agriculture has eliminated most occurrences. Some have been excavated for farm dams or filled and turned into pasture. The remaining occurrences are threatened by hydrological change, cattle grazing, weed invasion and altered fire regimes.

Status: Critically Endangered. An Interim Recovery Plan has been prepared and is being implemented by the Swan Region Threatened Flora and Communities Recovery Team. Two of the three remaining communities have been purchased and are now nature reserves. The remaining one is part of a 'Bush Forever' site within a major urban development.

Below Kings Mound Spring in Bullsbrook showing flooded gums, rushes and bracken ferns.

Photo – Val English/CALM

Aquatic root mat communities of caves of the Swan Coastal Plain

Description: In some caves at Yanchep, springs and pools support dense growths of roots (root mats), developed by trees growing on the surface above the cave. The root mats provide a constant and abundant primary food source for some of the richest aquatic cave communities known. The communities comprise a complete food web: the rootlets and their associated microflora provide the primary food source, and the invertebrate assemblages include root mat grazers, predators, parasites, detritivores and scavengers.

Distribution: Five caves in Yanchep National Park—YN99, Cabaret Cave, Carpark Cave, Twilight Cave and Water Cave (in the past Gilgie Cave also had root mats)—have streams or pools, fed by groundwater from the Gnangara Mound, that contain root mats from tuart trees (*Eucalyptus gomphocephala*).

Biological characteristics: The caves are considered to contain one ecological community because there is considerable overlap of animal species between the five caves, and water chemistry is very similar in all caves. Nevertheless, the faunal assemblages vary both in species composition and relative abundance of species. Aquatic cave animals at Yanchep include night fish, gilgies, leeches, microscopic worms, mites, snails, insects and crustaceans. Some species appear to be found nowhere else but in these cave streams, and some are confined to a single cave. About 100 animal species have been located in the six caves with the root mat communities. About a third of these are newly-discovered and still undescribed. Furthermore, at least six newly discovered species of crustaceans that occur in the community at Yanchep are relicts from when Australia was part of the supercontinent of Gondwana. The primary food source for the root mat community is the roots of mature tuart trees that extend into the caves, and probably the extensive fungal growth within the tissue of the rootlets.

Threats: The main source of water for the cave streams is groundwater emerging into the streambed within the cave, driven by the hydraulic head of the Gnangara Mound. The survival of a root mat community within any cave would be seriously threatened by drying of the stream in that cave. One occurrence, in Gilgie Cave, seems to have been destroyed after it dried out in the mid-1990s. The major immediate threat is the lowering of the groundwater table in the Gnangara Mound, due to utilisation from bores, drawdown by pine plantations and drought. Upstream of the caves, the level of the Gnangara Mound has dropped by up to 5 m since around 1976.

Below Root mat close-up showing new growth (white shoots), Cabaret Cave.

Photo – Michael James/CALM

Continuing groundwater extraction from the deep Yarragadee formation for the Perth water supply may contribute to additional lowering of the Gnangara Mound. Pollution and vandalism are also a concern, as is the possible destruction or death of the tuart trees that provide the root mats.

Status: Critically Endangered. The Department of Conservation and Land Management has prepared a recovery plan, and the Swan Region Threatened Flora and Communities Recovery Team is implementing it. Water levels in the pools are regularly monitored. In recent summers, many of the pools would have dried without artificial provision of water. This has been done by digging sumps that reach the declining groundwater, and by establishing a small pump and ball-cock mechanism to maintain water around the root mats. Currently, trials are taking place, with the support of the Water Corporation and Waters and Rivers Commission, to establish a longer-term,

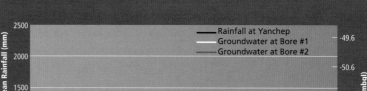
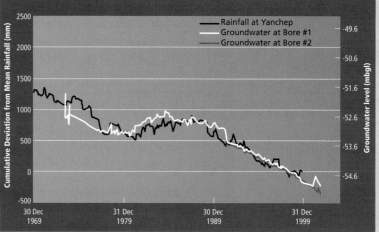

more natural and more robust water supply to caves with root mats, as well as to Crystal Cave, to conserve the Crystal Cave crangonyctoid (see p. 163). The only long-term solution is for groundwater levels in the area to return to higher levels. This will require removal of many of the pines and limiting water abstraction from the mound.

Top National park ranger Paul Tholen with remedial water pumping and trickle feed set up on a section of root mat to maintain water flow at Cabaret Cave.

Photo – Michael James/CALM

Above Calgardup Cave in the Leeuwin-Naturaliste Ridge.

Photo – Michael James/CALM

Aquatic root mat communities Nos 1–4 of the Leeuwin-Naturaliste Ridge

Description: On the Leeuwin-Naturaliste Ridge permanent streams occur below the surface and support dense root mats. The root mats, which provide a constant and abundant primary food source, sustain some of the richest faunal communities known from groundwater in caves anywhere in the world. The communities comprise a complete food web: the rootlets and their associated microflora provide the primary food source, and root mat grazers, predators, parasites, detritivores and scavengers complete the interactions.

Distribution: Calgardup, Easter, Kudjal Yolgah and Strongs caves, on the Leeuwin-Naturaliste Ridge, contain aquatic root mat communities. These caves occur within 20 km of the coastline on a Tamala (coastal) Limestone ridge that rises to 220 m above sea level. The four communities are considered to be distinct, and have been listed separately, but are treated together here.

Biological characteristics: Each of these caves is considered to contain a distinct community, as the species composition differs significantly in each. Root mats are produced by karri (*Eucalyptus diversicolor*) in Easter and Strongs caves, marri (*Corymbia calophylla*) in Calgardup Cave and karri and peppermint (*Agonis flexuosa*) in Kudjal Yolgah Cave. Aquatic cave animals in Leeuwin-Naturaliste caves include koonacs (*Cherax preissii*), mites, rotifers, microscopic worms, tardigrades, insects and crustaceans. Some species appear to be found nowhere else but these cave streams, and some are confined to a single cave. A total of 37 animal species (excluding nematodes and rotifers, for which the individual species have not yet been identified) have been located in the four caves that contain root mat communities. At least half of these are new to science. At least three amphipods, and the crustaceans that occur in the communities, are relicts from when Australia was part of the supercontinent of Gondwana, or even the earlier Pangaean period.

The rootlets contain extensive growths of microscopic fungi within their tissues that probably increase the nutritional value of the mats. The root mats, or their detritus, house more than 50% of the animals that occur in any one cave stream. The remainder occur in open water, and in the soils in the streambed. None of the Gondwanan relicts found in the caves appears to have drought resistant stages, indicating that they require permanent water for their survival.

Threats: Two of the caves are within, and two are outside, Leeuwin-Naturaliste National Park. The greatest concern is a decline in the regional water table from groundwater utilisation, leading to drying of the underground streams. Other possible threats include destruction of the trees or tree roots, pollution of the groundwater and invasion of exotic species such as yabbies (*Cherax destructor*).

Status: Critically Endangered. An interim recovery plan has been prepared and is about to be revised. It is being implemented by the Recovery Team for aquatic root mat communities Nos 1–4 of the Leeuwin-Naturaliste Ridge.

Camerons Cave troglobitic community

Description: Camerons Cave troglobitic community is known only from Camerons Cave near Exmouth. The community has a unique composition of troglobitic species, at least eight of which are known only from this location. The assemblage is related to those found in caves in Cape Range, but all species with congeneric members in caves in nearby Cape Range have, to date, proved to be distinct. The listed threatened species Camerons Cave millipede (*Stygiochiropus peculiaris*) and Camerons Cave pseudoscorpion (*Hyella* sp.) are endemic to Camerons Cave (see p. 148).

Distribution: Camerons Cave is located within Exmouth townsite.

Biological characteristics: The primary source of food for the Camerons Cave Troglobitic Community is tiny particles of, and dissolved sources of, organic carbon. This food is derived from outside the cave, and is transported primarily by the influx of water. Hence, the intensity, frequency and periodicity of rainfall will determine their energy recharge. The community relies on the humid conditions in Camerons Cave, created through contact with the water table, as well as specific surface conditions.

Threats: The cave is not protected by reservation, and the area around the cave is subject to various proposed developments. Threats include lowering of the water table, modification to the groundwater catchment, uncontrolled access by people, and pollution. Groundwater in the Exmouth area is used for human consumption and is managed by the Department of the Environment.

Status: Critically Endangered. A recovery plan has been prepared and is being implemented by the North West Cape Karst Management Advisory Committee. A locked gate has been installed on the cave entrance to limit access. The Exmouth Shire is aware of the cave and its values, and is working to ensure that land development does not interfere with it.

Above A troglobitic leafhopper (*Phaconeura* sp.) from the Cape Range area.

Photo – Douglas Elford /WA Museum

Below The entrance to Camerons Cave.

Photo – Bill Humphreys /WA Museum

Cape Range remipede community (Bundera Sinkhole)

Description: Bundera Sinkhole provides a tidally-influenced habitat, with water low in oxygen below a density-induced layer (thermo-halocline) separating brackish surface and deeper saline waters. This sinkhole is the only deep tidally-influenced (anchialine) system known in Australia. It has a surface opening about 1.7 km inland, and subsurface connections with seawater.

Distribution: Bundera Sinkhole is located on the western coastal plain of North West Cape peninsula.

Biological characteristics: Conditions in the cave are complex and easily disrupted. It supports a rich fauna composed primarily of crustaceans but including a blind fish. The community contains a number of species, occurring only below the thermo-halocline, that are unique to the site. These include the Cape Range remipede crustacean (*Lasionectes exleyi*), the Cape Range amphipod (*Liagoceradocus branchialis*) and the ostracod *Danielopolina kornickeri*, all species listed as threatened,

as are the calanoid copepod crustaceans *Bunderia misophaga* and *Stygocyclopia australis*, the misophrioid crustacean *Speleophria bunderae* and the Cape Range bristle worm (*Prionospia thalanji*). Other crustaceans in the sinkhole include a shrimp, other ostracods, gammarid amphipods and diverse copepods. In addition, the adjacent aquifer includes cirolanid isopods, a second species of atyid shrimp, thermosbaenaceans, diverse amphipods and the blind cave eel (*Ophisternon candidum*) (see p. 139).

Most of the remarkable invertebrate animals of Bundera Sinkhole and its adjacent aquifer are relicts from the Tethyan Sea, which in Mesozoic times separated Laurasia from Gondwana following the break-up of the Pangaean supercontinent. The cave is home to the same groups of crustaceans that have been recorded from tidally-influenced caves in the Bahamas, the Yucatan Peninsula of Mexico, and Cuba.

Threats: The cave is on Commonwealth land controlled by the Department of Defence as a military exercise area and bombing range. A management agreement between that Department and the Department of Conservation and Land Management provides some protection. Threats include pollution of the cave's water, SCUBA diving (which mixes the layers in the water) and the introduction of exotic species. These threats are compounded because of the single occurrence, and the cave's location near a vehicle track in a remote part of the peninsula.

Status: Critically Endangered. An interim recovery plan has been prepared and is being implemented by the North West Cape Karst Management Advisory Committee.

Below A blind shrimp (*Stygiocaris stylifera*).

Photo – Douglas Elford /WA Museum

Left *Pygolabis humphreysi,* an invertebrate species from the Ethel Gorge Aquifer.

Photo – Buz Wilson /Australian Museum

Ethel Gorge Aquifer stygobiont community

Description: This stygofaunal community occupies a freshwater aquifer, and includes species of amphipods, isopods, copepods and ostracods—all crustaceans—as well as flatworms.

Distribution: The aquifer is in and near Ethel Gorge, in the Fortescue River valley east of Newman.

Biological characteristics: After initial sampling of the aquifer by the WA Museum, 15 new species of amphipods in the new genus *Chydaekata* were described, 13 of which were known only from the Ethel Gorge aquifer. Six of these were collected from a single bore. In addition, five species of ostracods in four genera have been described from the aquifer, none of which are currently known from any other locality.

Threats: The major threat is groundwater drawdown due to mine dewatering. In 2000, BHP Iron Ore sought approval to mine an ore body alongside the aquifer, a process that would require dewatering the aquifer. Because this might, on knowledge available at the time, have led to the extinction of one or more of the species in this ecological community, the Department of Conservation and Land Management interacted with BHP and the Environmental Protection Authority to obtain additional information on the amphipod species and their distribution. BHP decided not to proceed with the mine until the situation was clarified. Genetic research showed that 11 of the 15 presumed species were probably actually two morphologically very variable species (the other four could not be relocated). The two species identified by the genetic research also occur outside the area of water drawdown. It was concluded that it was unlikely that the proposed mine would lead to extinction of any biological species, but it would still impact on the ecological community.

Status: Endangered.

Species-rich faunal community of the intertidal mudflats of Rocbuck Bay

Description: Roebuck Bay is a sheltered marine embayment with large flats composed of sediments of carbonate origin exposed at low tide. It has relatively little fresh water input and slow tidal flows.

Distribution: Roebuck Bay, near Broome.

Biological characteristics: Roebuck Bay supports a unique species-rich assemblage of marine creatures (particularly invertebrates that live in the sediments on the floor of the bay). Other apparently-similar marine embayments in the tropics, such as King Sound and Darwin Harbour in Australia, are known to have quite different and less diverse assemblages of marine invertebrates. Roebuck Bay also provides habitat for a high diversity and abundance of transequatorial migratory shorebirds. It is a Wetland of International Importance and is recognised as one of the world's most important sites for migratory waders. Up to 150,000 waders use the bay during their migration between Australia and the northern hemisphere. The species-rich faunal community of the intertidal mudflats of Roebuck Bay occupies 28,000 ha alongside the Broome townsite.

Threats: Along much of the northern shores of Roebuck Bay, high-tide shorebird roosts are vulnerable to disturbance from off-road vehicles and passive pedestrian traffic, and there is an increasing amount of tourism pressure on the southern parts of the bay, around Bush Point and Sandy Point. Patterns of roost choice by red knots (*Calidris canutus*) in Roebuck Bay are determined by a combination of local topography, disturbance and microclimate. Birds roost on the highly-disturbed northern beaches of Roebuck Bay during tide conditions when all potential alternative roosts have a dangerously hot microclimate. In addition, it is beneficial for shorebirds to roost close to their feeding grounds. Where disturbance is significant, birds will expend energy travelling between feeding grounds and alternative roosting sites. Minimum energy reserves are critical for birds to complete transcontinental migration.

There is a risk of industrial and urban wastes entering the intertidal ecosystem of Roebuck Bay from spillage, run off, sewerage, groundwater flows and airborne or mechanical means. In particular, there are proposals to grow

Right So far, more than 100 species of bivalves have been found in Roebuck Bay. This handful includes at least six different species.

Below The intertidal mudflats of Roebuck Bay are known for their amazing diversity of invertebrates, including crabs. Here a land crab (*Neosarmatium meinerti*) feeds on a flame fiddler crab (*Uca flammula*).

Photos – Jan van de Kam

cotton under irrigation along the coastal plain east and south of Roebuck Bay, using either groundwater or a dam on the Fitzroy River. Cotton farming is notorious for its dependence on high levels of chemicals for pest control and fertilisers. Even new pest-resistant cotton strains will not significantly reduce chemical use. Mining activities, dredging and reclamation of mudflats are also potential threats, and excessive pumping of groundwater from the shallow aquifers of the hinterland may have a detrimental effect on the surface water expressions of the Roebuck Plains.

Status: Vulnerable. Considerable research into this community has been conducted as a joint project between the Netherlands Institute for Sea Research, the Department of Conservation and Land Management and Curtin University of Technology. The area is a proposed marine conservation reserve.

Above The intertidal mudflats of Roebuck Bay are recognised as one of the world's most important sites for migratory waders. Eastern curlews, bar-tailed godwits and silver gulls have landed in the waterline east of the Broome Bird Observatory. Smaller shorebird species, such as red knots and curlew sandpipers, are alighting.

Photo – Jan van de Kam

Above Three Springs mound springs.

Photo – Sheila Hamilton-Brown /CALM

Assemblages of the organic mound springs of the Three Springs region

Description: The habitat of this community is characterised by continuous discharge of groundwater in raised areas of peat. The peat and its surrounds provide a stable, permanently moist series of microhabitats. While the flora assemblages associated with these mound springs comprise species that are common in streamside and riverside areas, and widespread throughout the region, the invertebrate fauna assemblages include species that are rare or absent in other types of aquatic habitat in the Wheatbelt.

Distribution: There were originally 22 organic mound springs occurring from Three Springs north to near Eneabba.

Biological characteristics: There are considerable differences in invertebrate fauna assemblages between these sites, and all are associated with a rich and healthy fauna.

Threats: Only half of the original number of mound springs now exist. All but one are on private or unallocated Crown land. The mound spring on a conservation reserve is the smallest, contains only a third of the plant species and is an outlier that is not representative of the others. Current threats include hydrological change, destruction, grazing, fire and weed invasion.

Status: Endangered.

Depot Springs stygofauna community

Description: The Depot Springs groundwater calcrete contains a unique assemblage of stygofaunal species, dominated by water beetles.

Distribution: The calcrete occurs between Sandstone and Leinster.

Biological characteristics: The assemblage includes two dytiscid beetle species that are known only from the Depot Springs calcrete: *Nirridessus fridaywellensis* and *Nirripirti hinzeae*. The latter species is known only from Friday Well.

Threats: The community appears to have a highly localised distribution and is at risk from groundwater abstraction for mining purposes.

Status: Vulnerable.

References

Black, S., Burbidge, A., Brooks, D., Green, P., Humphreys, W.F., Kendrick, P., Myers, D., Shepherd, R. and Wann, J. (2001.) Cape Range remipede community (Bundera Sinkhole) Interim Recovery Plan 2000–2003. Interim Recovery Plan No. 75. Department of Conservation and Land Management, Wanneroo.

Black, S., Burbidge, A., Brooks, D., Green, P., Humphreys, W.F., Kendrick, P., Myers, D., Shepherd, R. and Wann, J. (2001.) Camerons Cave troglobitic community, Camerons Cave millipede and Camerons Cave pseudoscorpion Interim Recovery Plan 2000–2003. Interim Recovery Plan No. 76. Department of Conservation and Land Management, Wanneroo.

English, V. and Blyth, J. (1999). Development and application of procedures to identify and conserve threatened ecological communities in the South-west Botanical Province of Western Australia. *Pacific Conservation Biology* 5, 124–138.

English, V. and Blyth, J. (2000). Aquatic root mat communities Nos 1–4 of the Leeuwin-Naturaliste Ridge Interim Recovery Plan 2000–2003. Interim Recovery Plan No. 53. Department of Conservation and Land Management, Wanneroo.

English, V. and Blyth, J. (2000). Assemblages of Organic Mound (Tumulus) Springs of the Swan Coastal Plain Interim Recovery Plan 2000–2003. Interim Recovery Plan No. 56. Department of Conservation and Land Management, Wanneroo.

English, V., Jasinska, E. and Blyth, J. (2003). Aquatic root mat community of caves of the Swan Coastal Plain, and the Crystal Cave Crangonyctoid Interim Recovery Plan 2003–2008. Interim Recovery Plan No. 117. Department of Conservation and Land Management, Wanneroo.

Rogers, D., Piersma, T., Lavaleye, M., Pearson, G. and de Goeij, P. (2003). *Life along land's edge, wildlife on the shores of Roebuck Bay, Broome.* Department of Conservation and Land Management, Kensington.

chapter 11

A vision for the future

by Andrew Burbidge and Keiran McNamara

Western Australia's native animals and plants have not fared well since settlement by Europeans. In the past 175 years, 17 animal species have been listed as extinct (see Chapter 1), however, the real number of extinctions is probably very much higher, as many species of invertebrates doubtless disappeared undetected when their habitat was destroyed. Officially, 185 species and subspecies are listed as threatened with extinction in WA. Again, this figure probably represents only the tip of the iceberg due to lack of knowledge (see Chapter 9).

Reasons for the loss of biodiversity were discussed in Chapter 2, where the 'four horsemen of the apocalypse'– habitat destruction, habitat degradation, invasive species and climate change– were identified and discussed. Current conservation activities were overviewed in Chapter 3.

If we do nothing, or even if we maintain current programs, more species will be lost. Fortunately, today's society is very different from that of pioneering days, and many people now understand and support the conservation of our biodiversity, and want governments and the broader community to commit the necessary resources to prevent any further losses.

Resources are not unlimited and conservation activity, like any other endeavour, must be carried out efficiently and effectively. Conservation has to compete for resources with other important activities—such as health, education and welfare. Conservation planning must be built on the principle that prevention is better than cure: ensuring that species and communities never get to a crisis level, where they must be listed as threatened and intensively managed, is much better than the expensive 'cure' that follows once a recovery plan becomes necessary.

Above Information collected during a major study of the Archipelago of the Recherche will help in the planning of a proposed marine conservation reserve in the area.

Photo – Justin McDonald

Opposite Because of their isolation, the Rowley Shoals provide one of the best chances anywhere in the world to preserve a pristine coral reef system. Extensions to the Rowley Shoals Marine Park are planned to better protect this magnificent area.

Photo – Ann Storrie

Below Aerial view of part of the Wheatbelt where very few, mainly small, areas of remnant vegetation remain, many of which are threatened by increasing salinity.

Photo – Ken Wallace/CALM

The vision

Above A young female Gilbert's potoroo from a colony discovered in February 2003 by a team comprising CALM staff, Gilbert's Potoroo Action Group members and other volunteers. Such partnerships are vital for effective conservation programs.

Photo – Helen Crisp/Gilbert's Potoroo Action Group

We look forward to a Western Australia where:

- there is no extinction of native species through human action or inaction;

- WA's animals, plants and micro-organisms persist in viable populations where they can continue to adapt and evolve;

- people recognise that conservation of natural biodiversity is central to their quality of life and critical to meeting their fundamental needs—so that biodiversity conservation is regarded as an essential service of equal value to programs for health, education and law and order, and there is strong support for biodiversity conservation by the vast majority of people living in WA;

- the hidden subsidies currently borne by the environment are incorporated into the cost of commodities, food, fibre and water, and economic evaluations and environmental impact assessment fully account for the cost of the 'free' ecological services provided by biodiversity;

- 'greenhouse gas' emissions leading to climate change are controlled and actions are underway to prevent species extinction and ecological community destruction from that cause;

- there is a comprehensive, adequate and representative (CAR) system of terrestrial and marine conservation reserves, with further protected areas to conserve threatened species and threatened ecological communities, as well as features of scientific interest and local importance, noting that many other reserves are needed across the landscape to conserve local examples of ecosystems, and to allow species to move between reserves, even if those reserves contain nothing especially rare;

- freehold and leasehold land is managed sustainably with biodiversity, water and land conservation as one major objective;

- threatened species and threatened ecological communities are identified, legally listed and protected, and conserved according to published recovery plans;

- there is a well-resourced, publicly-supported government agency managing the State's biodiversity and conservation reserves on behalf of the people of WA;

- there are effective partnerships between government and the community, promoting and conducting conservation activities;

- conservation organisations have engaged managers of production systems and helped them to develop sustainable, profitable systems that are environmentally sensitive; and

- we live in a truly sustainable society where people live in harmony with the natural world.

Significant changes are needed in the way our society functions, and new programs and resources will be necessary before this vision can be achieved. This is not to say that nothing has been achieved so far. Much has been accomplished, but much remains to be done. The approach must be broad, with many aspects of human endeavour requiring modification, and research programs being needed to develop and apply new technologies.

Our vision cannot be achieved by government and its agencies alone—all of society must contribute. In the rest of this chapter we will discuss some of the major biodiversity conservation issues that need to be addressed in the immediate future.

Law and economics

Legislation

Good legislation is fundamental to the development of biodiversity conservation programs. Each of the three levels of government has a role to play.

At the federal level, the *Environment Protection and Biodiversity Conservation Act 1999* (the EPBC Act) applies to Commonwealth-owned land, external territories, Australian territorial seas and matters that are the Constitutional responsibility of the Commonwealth, such as external affairs and regulation of trade. The EPBC Act takes precedence over State and local laws on some aspects of environmental impact assessment and control, particularly in relation to threatened species and threatened ecological communities. The Commonwealth agency responsible for implementing the EPBC Act, the Department of the Environment and Heritage, seeks to work closely with State and Territory conservation departments to ensure that the Act's purposes are achieved. Primarily through the Natural Heritage Trust, the Commonwealth provides funds to support conservation projects.

Most biodiversity conservation in Australia is carried out at the State and local level. Land use and land management are basic to good conservation outcomes, and under the Constitution these are a State government responsibility. Thus, Australian national parks are declared and managed by State governments, and not by the federal government as is the case in the United States of America. In WA, two major Acts guide biodiversity conservation practices. The *Conservation and Land Management Act 1984* deals with the creation and management of national parks, nature reserves, marine parks, marine nature reserves, conservation parks, State forests and other conservation land categories. The *Wildlife Conservation Act 1950* protects the State's indigenous plants and animals, and lays down the basis for their management. Many other Acts affect biodiversity, and it is important that principles established in biodiversity conservation legislation are complemented and not abrogated by other legislation.

While it was pioneering legislation in the 1950s, 1960s and 1970s, the Wildlife Conservation Act is now out of date, and the State government is committed to introducing a modern Biodiversity Conservation Bill in the near future. The Bill will include provisions dealing with listing and conservation of threatened species and threatened ecological communities, the development and implementation of recovery plans for listed species and communities, and the management of threatening processes. Other important proposed provisions include the identification and protection of critical habitat (habitat critical to the survival of a species or ecological community), the development of voluntary conservation agreements, new conservation covenanting arrangements, the listing of 'key threatening processes', special controls for biological threats and the development of bioregional plans. The Bill will also promote the provision of financial assistance and incentives to encourage the management and restoration of biodiversity.

Above The Commonwealth Government's Natural Heritage Trust provides funds to support conservation projects such as the preparation of a recovery plan for the kingo, or red-tailed phascogale.

Photo – Babs and Bert Wells /CALM

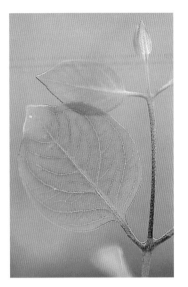

Above Oil mallee seedling.

Photo – Marie Lochman /CALM

Below Oil mallees are being used in plantings on farmland to generate income while combating the effects of salinity.

Photo – Jiri Lochman/CALM

Economics and ecology

Although the English words for these two vital areas of human endeavour arise from the same Greek word (*oikos*, meaning 'house'), the two activities have largely ignored each other. Economic modelling and analysis have not adequately included what economists term 'externalities'—matters such as the effects of an activity on biodiversity and the environment (for example, release of greenhouse gases, pollution of waterways or destruction of native vegetation), and the provision of 'free services' by the environment (for example, clean air and water, and pollination of crops by native animals). Thus, there is a failure in the marketplace to value environmental services and cost environmental damage. This system of accounting has led to many of the environmental problems that we see today, and there is an urgent need for a paradigm shift in the way in which society values the environment.

Without such changes, it is difficult to see how true ecologically sustainable development can ever occur. The extinction of species and the large number of threatened species in Western Australia is, to a significant degree, the result of the failure of society to properly value biodiversity.

Sudden changes in the way markets are regulated, and the way services are valued, are not possible without major impacts on economic activity. However, a gradual change is both possible and necessary if we are to sustain the environment on which biodiversity and people depend. Already, work is well underway in Australia to develop public policy options in areas such as environmental water flows and water property rights, environmental services in plantation forestry and fostering new farming systems.

The conservation reserves system

Terrestrial conservation reserves

The development of a comprehensive, adequate and representative (CAR) biodiversity conservation reserves system was discussed in Chapter 3. Data presented there (p. 27) show that one terrestrial bioregion has no reserves at all, and 14 of WA's 26 bioregions have less than 10% of their area reserved. In some of the bioregions with a relatively high proportion of land reserved, the reserves include large areas of unproductive land, and do not include any of the more productive areas—thus, they are not 'comprehensive'.

Clearly, there are significant gaps in WA's conservation reserves system, and it is a long way short of meeting CAR criteria. Strategies to create additional reserves and moves towards fulfilling CAR criteria include:

- the reservation proposals in the Government's 'Protecting our old-growth forests' policy;

- purchase of pastoral leases, particularly under the Gascoyne-Murchison Strategy;

- identification of whole and part pastoral leases that are required for conservation purposes, and hence that should not be renewed when leases expire in 2015;

- purchase of freehold remnant vegetation and wetlands in the south-west agricultural zone and on the Swan Coastal Plain, with priority being given to threatened species and ecosystems, as well as bushland contributing to State Salinity Strategy outcomes;

- the reservation proposals for the Perth Metropolitan Region in 'Bush Forever';

- pursuit of Environment Protection Authority Conservation Through Reserves Committee recommendations that remain unimplemented and that have not been superceded; and

- pursuit of areas identified in State planning strategies, Department of Conservation and Land Management (CALM) regional and area management plans, and as a result of biological surveys.

Biogeographic survey is basic to the development of a CAR reserves system, and identification of elements of biodiversity in need of special attention. WA leads the nation in the design, implementation and analysis of regional terrestrial biogeographic surveys, with CALM's Science Division taking the lead role. Major, published surveys of the central deserts, Nullarbor Plain, eastern Goldfields, Kimberley rainforests, Kimberley islands, and southern Carnarvon Basin have contributed greatly to progress towards a CAR reserves system, and to conservation priorities and practices. A detailed survey of the WA Wheatbelt is complete and a survey of the Pilbara region is underway.

One important aspect of these surveys is that they are site based, quantitative and repeatable. It will be possible for future biogeographers to return to exactly the same survey sites in future years, knowing precisely which data-collecting techniques were used. With a State as large as WA, current resources do not permit repeat surveys at appropriate intervals. With additional resources, CALM will be able to complete the first set of regional surveys by 2020, and resurvey each region every 20 years.

Above Members of the local catchment group view the tumulus mound springs at Bullsbrook. In 2001, a 9 ha block was purchased to protect this Critically Endangered threatened ecological community.

Photo – Val English/CALM

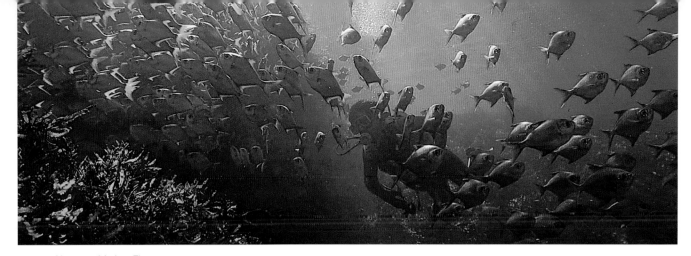

Above and below The
proposed 'Capes' or
Geographe Bay/Leeuwin-
Naturaliste/Hardy Inlet
marine conservation reserve.

Photos – Mick Eidam

Another basic requirement for
conservation reserves is good
management. Reserve managers currently
lack good information on how to
manage natural areas, including
remnants, for the best conservation
outcomes, and additional research is
badly needed. CALM aims to increase
long-term research into ecosystem and
species dynamics within conservation
reserves, and to boost reserve
management capacity.

The developing law and policy
regarding Native Title means that
traditional owners and CALM will jointly
manage many of the existing and future
conservation reserves. The government
proposes to amend the Conservation and
Land Management Act to provide for the
establishment of boards of management
with a majority of traditional owners,
and to provide title in perpetuity or
management orders for parks and reserves
to be held by Aboriginal corporate bodies
representing traditional owners.

Marine conservation reserves

The creation of a CAR reserves
system in marine and estuarine
environments is at an early stage. Before
a marine conservation reserve can be
declared, there has to be considerable
consultation with affected people and
industries, and the preparation of a draft
management plan. Historically, public
access to marine areas has been largely
unfettered and, apart from fisheries
regulations, there have been few
restrictions on activities. This is a very
different situation from that on land,
where boundaries and ownership rules
have been long established. Nevertheless,
most people are coming to accept that
biodiversity conservation and resource
management are needed as much in

marine areas as on the land. With such
acceptance, the creation of marine
conservation reserves will become
smoother. The popularity and recognition
of the Ningaloo Marine Park bodes well
for the further development of the system.

The 1994 Report of the Marine
Parks and Reserves Selection Working
Group identified about 70 areas as
candidate marine conservation reserves.
In June 2003, there were only six marine
parks and one marine conservation
reserve; thus the system is far from
complete. In 1998, the State government
published *New Horizons: the way ahead
in marine conservation and management*,
its policy for the creation of new marine
conservation reserves. This policy
committed to the development of a
world-class multiple-use marine
conservation reserve system, to be
achieved by the progressive detailed
evaluation of the areas in the Working
Group Report.

Data in Chapter 3 (p. 28) shows
that 10 of the 18 IMCRA regions in WA
have no marine conservation reserves,
while others have a very small proportion
of their area reserved.

The Jurien Bay Marine Park, the
State's seventh marine park, was declared
in August 2003. Several marine
conservation reserves are in the pipeline,
at Montebello-Barrow Islands, Dampier
Archipelago-Cape Preston, Geographe
Bay/Leeuwin-Naturaliste/Cape Hardy Inlet
area and Walpole-Nornalup inlets.
Extensions to the Ningaloo, Rowley
Shoals, Shark Bay and Shoalwater Islands
marine parks are also planned. CALM's
goal is to complete reservation or further
investigation of all the reserves proposed
by the Working Group Report by 2020.

Conservation outside reserves

If we are to achieve biodiversity conservation goals, conservation cannot be restricted to national parks and other conservation reserves, as vital as they are. CALM is increasing its commitment to helping landowners conserve native plants and animals on their land and to supporting community groups throughout the State, including those working to conserve marine environments. Supporting regional and local communities is also a major focus of the Natural Heritage Trust.

The land and its resources need to be managed on a landscape scale, integrating primary production and other land-use activities with conservation. Providing people with information about land and biodiversity management, and developing cooperative programs on regional scales, is becoming a significant Departmental function. Basic to this process in the south-west agricultural areas are programs to combat land degradation, especially that due to increasing salinisation and waterlogging.

The recently-completed biogeographic survey of the Wheatbelt, conducted by CALM's Science Division and collaborators, has shown that salinity is a major threat to biodiversity, as well as to farming. Departmental scientists expect that, under a 'do nothing' scenario, the remaining populations of 450 species of flowering plants, more than 100 species of ground-dwelling spiders and scorpions, and more than 300 species of freshwater invertebrates will become so fragmented and reduced that they are likely to become extinct within a century (although a few of the freshwater species may survive in farm dams or in adjacent districts). Many other species of invertebrate animals not included in the survey, such as terrestrial insects, may also disappear.

Of the 18.8 million hectares of agricultural land in the south-west of WA, 4.1% is currently affected by salinity. The Department of Agriculture estimates that 23.9% of agricultural land is potentially at risk. Of the conservation estate managed by CALM, 196,500 ha are currently salt affected, with 764,000 ha potentially at risk. These dramatic figures show what an enormous threat salinity is, and how important it is to take steps to reduce its impact.

The Department's developing approach to salinity and related hydrological issues includes the following.

- A Crown reserves program that aims to contribute to the protection and restoration of high-value wetlands and natural vegetation, maintain natural diversity within the region, and to improve the management and protection of native vegetation remnants, so that their long-term contribution to salinity control is maintained and, where practicable, improved.

- A natural diversity recovery catchments program that aims to develop and implement a coordinated wetlands and natural diversity recovery program targeting key catchments, to ensure that critical and regionally significant natural areas, particularly wetlands, are protected in perpetuity.

- A 'Land for Wildlife' program that aims to encourage and assist landholders to provide habitats for wildlife on their property.

- A biological survey program that was undertaken in the agricultural zone, with an emphasis on low-lying areas vulnerable to salinity. The survey, completed in 2003, aimed to identify and prioritise potential recovery catchments, provide a regional perspective on nature conservation

Above Animals such as the numbat are relatively immobile and cannot cope with the effects of salinity.

Photo – Babs and Bert Wells /CALM

Above Andrew Storey sampling for aquatic invertebrates in the Lake Muir recovery catchment during a comprehensive four-year biological survey of the agricultural zone undertaken as part of the State Salinity Strategy.

Photo – Greg Keighery /CALM

priorities, provide baseline data and a regional framework for future monitoring, and provide other information.

- A wetland monitoring program that aims to determine long-term trends in natural diversity within wetlands, to provide a sound basis for corrective action.

- An oil mallee program that aims to establish a new commercial industry in rural areas that has multiple benefits to the State. Such an industry would improve recharge control in the low rainfall zone, by providing a commercially viable option for increasing water use.

- A project that aims to develop a procedure that systematically analyses native plants of low/medium rainfall areas to identify their best prospects for development as tree and shrub crops. The initial screening phase of this program is nearly complete, and more intensive development work will shortly commence. The program will also broaden its scope to include, in the revegetation design phase, options for tackling a wider range of threats to biodiversity than salinity.

Controlling threatening processes

Salinity (and associated land degradation) is but one of many threatening processes that need to be controlled if species and ecological communities are to persist. Paramount among threats to WA's animals are habitat destruction through land clearing, habitat degradation and invasive species (see Chapter 2).

Land clearing

Broadscale land clearing for agriculture is largely a thing of the past in WA. However, animal habitat is still being destroyed through, for example, the removal of trees for exotic tree plantations and vineyards, because farming practices are becoming more mechanised, and for urban development and infrastructure. Around Perth, banksia woodlands are still being cleared for housing. Near the south-west coastline, new housing developments are also destroying habitat. For example, numerous small developments are gradually reducing the peppermint (*Agonis flexuosa*) woodland between Bunbury and Augusta. This is the prime habitat of the threatened western ringtail possum (*Pseudocheirus occidentalis*). Ways of balancing housing development with biodiversity conservation requirements are necessary—at present the way in which urban planning and environment impact assessments are conducted is inadequate for dealing with cumulative impacts, and may lead to a 'death by a thousand cuts' of species and communities.

As the fundamental result of habitat destruction and degradation is insufficient key ecological resources (food, water, shelter, access to mates) to support viable populations of native plants and animals, an important operational response is to increase the area of effective habitat. The most obvious means is through revegetation, however, in the marine environment artificial reefs are also an example.

Fire

Changed fire regimes can be a major contributor to land degradation. As discussed in Chapter 3, knowledge of the effects of fire on biodiversity, while it is expanding rapidly, is still limited. This is partly because of the enormous variability in fire regimes—a term that embraces fire intensity, size, season and frequency—and partly because of the large variations in climate and vegetation between the many bioregions of WA. Fire is an integral part of most WA ecosystems, and fire management is a significant component of CALM's annual expenditure. On lands managed by CALM, the aim is to manage fire to conserve biodiversity, and to ensure an acceptable level of protection to human life and property.

A symposium on fire in south-west ecosystems, attended by some 350 fire scientists, fire managers, academics, volunteer conservationists, volunteer firefighters and interested community members, was held in Perth in April 2002. The scientific proceedings have been published. Notwithstanding the improvement in knowledge of fire and biodiversity in recent years, it is clear that

Below The noisy scrub-bird prefers long-unburnt vegetation. It has poor dispersal ability so is favoured by small, patchy fires rather than large intense fires.

Photo – Babs and Bert Wells/CALM

Above A mass-collapse site in jarrah forest, caused by the effects of dieback disease.

Photo – Marie Lochman /CALM

additional, long-term research into the relationship between fire and natural ecosystems is greatly needed in all the major bioregions in the State, to allow fire managers to adapt their use and control of fire on an ongoing basis. Such studies must take account of the needs of the rarer components of ecosystems, especially threatened species and ecological communities. As fire research is so important, it should be funded through universities and other research institutions, as well as from within government.

The Department's goal for 2020 is to have sufficient knowledge of fire ecology and fire behaviour for all major bioregions, to conserve biodiversity, while at the same time ensuring protection of human life and property within and adjacent to the land it manages.

Weeds

As discussed in Chapter 2, many of the alien plants established in WA are degrading natural ecosystems and landscapes. Some weed species are particularly widespread and virulent, and require special attention. Priorities vary between regions.

Currently, the size and complexity of managing weeds means that CALM is not able to tackle more than a few of the many weeds threatening WA's biodiversity. The Department's goal for 2020 is to have established a specialist weed control group that will assist regional staff to manage weeds, both within protected areas and where weeds are affecting threatened species, as well as to conduct and support research into effective control, including biological control, of priority weeds.

Preventing the introduction of new weeds is very important, and good screening systems are vital. Developing new industries based on local species will also help to obviate the need for introductions.

Dieback and other diseases

Phytophthora-caused 'dieback disease' (which actually causes the death of many native plants) is a major cause of land degradation in the south-west of WA, particularly in higher rainfall areas (see Chapter 2). Many species of native plants are killed by *Phytophthora cinnamomi*, the main pathogen involved, and some are threatened with extinction due to its effects. Around 60% of the mountain shrublands and banksia and mallee woodlands of the 116,000 ha Stirling Range National Park have been infested, as are perhaps 70% of the seasonally-inundated banksia woodlands in the Shannon and D'Entrecasteaux national parks. As many as 2000 of the estimated 9000 native plant species in the south-west of Western Australia are susceptible to *P. cinnamomi* root rot disease. Animals are indirectly affected through habitat degradation—species-rich areas of kwongan, for example, may be turned into species-poor sedgelands.

While CALM scientists have shown that spraying with phosphite can limit spread of the disease in natural ecosystems, there is no available technique to eliminate the organism once it establishes in the wild. Thus hygiene procedures, aimed at preventing introduction of *Phytophthora* species to uninfected areas, are vital. Further research, aimed at finding a method to eradicate *Phytophthora* in natural ecosystems, is very important. The Department's goal for 2020 is to find and implement a method for controlling this devastating disease. More information can be found at http://www.naturebase. net/projects/dieback_splash.html.

Knowledge of diseases affecting native wildlife, especially threatened animals, is poor. It is essential to ensure that threatened animals being translocated to new sites are disease free. Already, some translocations have been held up because of the presence of disease, for example, the wart-like disease found in western barred bandicoots from Shark Bay islands (see p. 59). The Department needs to work closely with wildlife disease scientists, from universities, zoos and elsewhere, to manage wildlife disease issues.

Western Shield

Western Shield, CALM's major introduced predator control and native wildlife recovery project, was discussed in Chapter 3. An external, independent review in 2003 stated that *Western Shield* was a world-class predator threat abatement program, strategically targeted at the recovery of a wide range of threatened wildlife species. The reviewers made several recommendations for its future improvement. These included expanding feral cat research and management, better defining the scope of the project, improving its management structure, promoting community involvement and improving monitoring. The report also discussed various more detailed aspects of *Western Shield*, including captive breeding, marooning and the role of islands, moving outside the south-west of the State, dingoes, public relations, and publication and communication of results.

The Department is committed to implementing the review's recommendations. The goal for 2020 is to be able to control feral cats effectively over large areas of the State and to extend *Western Shield* to conservation reserves in all regions of the State.

Control of introduced animals not included in *Western Shield*

Western Shield focuses on introduced mammal predators. As discussed in Chapter 2, other introduced animals are causing land degradation, including goats, donkeys, rabbits and wild cattle. Other introduced species such as honey bees, the predatory snail *Oxychilus* and the cane toad are affecting, or will affect, WA's biodiversity. Control of these species will benefit many threatened species, and prevent others from becoming threatened. The Department's goal is to create a specialist introduced animal control unit to carry out and coordinate the necessary control work. Animals that threaten WA's biodiversity include native species that have been introduced from other parts of Australia, such as the eastern long-billed corella (*Cacatua tenuirostris*) and rainbow lorikeet (*Trichoglossus haematodus*), which have become established in the Perth area and will probably extend their range unless controlled.

Introduced animals are also a major issue in the marine environment. Already many animals have been introduced to Australian coastal seas, often arriving in ships' ballast water. The Department will work with other agencies to try to prevent further introductions and to control those animals already established.

Conservation of threatened species not included in *Western Shield*

As important as it is, *Western Shield* covers only a relatively small proportion of WA's threatened animals. The approach to the conservation of these species has primarily been through the development and implementation of recovery plans. Because there are so many threatened species in WA, the concept of producing 'interim recovery plans' was developed. These summarise what is known about the species, prescribe urgent conservation actions to prevent its imminent extinction and propose further research necessary to produce detailed, full recovery plans. CALM celebrated the production of the

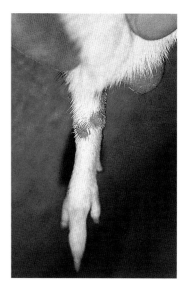

Above An example of the wart-like syndrome on the leg of a marl.

Photo – Stephanie Hill

Above The Endangered Stirling Range *moggridgea* is one of the species that could disappear as a result of climate change.

Photo – Jiri Lochman/CALM

100th interim recovery plan in March 2002, and by the end of 2003 there were 152 interim recovery plans for threatened animals, plants and ecological communities. Of these, nine are for animals and five are for threatened faunal communities. Most interim recovery plans have been for Critically Endangered species and ecological communities. The Commonwealth, through the Natural Heritage Trust, has assisted the preparation of recovery plans, as the EPBC Act (see p. 183) requires preparation of recovery plans for nationally-listed species within certain timeframes. However, the Commonwealth has no such requirement for the implementation of recovery plans adopted under their legislation.

As most WA-produced interim recovery plans have been for Critically Endangered taxa (species, subspecies or varieties), few Endangered taxa and very few Vulnerable taxa have recovery plans. As more recovery plans are finalised, it is becoming harder to implement them without additional staff and funds. As knowledge improves, more species and ecological communities will be listed. The goal of having recovery plans for all Critically Endangered and Endangered species and ecological communities by 2010, and being able to implement them effectively, can be achieved only by allocating new resources to this important work.

Climate change

It is now widely accepted that the global climate will change significantly in the next few decades, due to increasing levels of 'greenhouse gases' in the atmosphere. Predictions based on models, developed by CSIRO and others, suggest that the climate in the south-west of WA will become warmer and that rainfall will decrease. Elsewhere in the State, changes are also likely, for example, the inland Pilbara will become an even hotter place than it is now.

These changes will affect the distribution and abundance of animals. We know, for example, that mammal distribution in Australia is significantly affected by rainfall, and that drier areas have suffered much greater rates of extinction and decline than wetter ones. Thus, threatened mammals that inhabit areas where the rainfall decreases are likely to disappear. Some species are restricted to special habitats that may disappear. Several invertebrate animals occur only in small moist areas in high rainfall zones in the south-west corner of WA—these species are particularly at risk from climate change. Many wetland habitats will disappear. Stygofaunal species that occur in groundwater may also be affected, as groundwater levels drop and abstraction increases. Even marine ecosystems will be affected— already coral bleaching caused by high sea temperatures is occurring with increasing frequency in WA's northern waters. The increased pressure placed on animals due to climate change may be sufficient to tip many already-threatened species over the edge.

We need to better educate people about the possible effects of climate change on WA's biodiversity. We must incorporate climate change predictions into planning, for example, of the conservation reserves system. Corridors may allow some animals to move as the climate changes. We require new research to be able to better predict the outcomes of climate change. We need to review the distribution of threatened species in relation to climate change and, where possible, develop revised recovery plans. This is another area where CALM needs additional specialist staff.

Working together

Cooperative programs

Biodiversity conservation is everyone's responsibility. One government department, while taking the lead, can do only so much. Increasingly, private organisations are becoming involved in conservation programs. The Australian Wildlife Conservancy and the Australian Bush Heritage Fund, for example, are purchasing land to manage for conservation purposes, and in some areas predator control and species translocations are significantly benefiting threatened species. CALM has developed, and is developing, partnerships with many individuals and other organisations. Some of these are:

- individual landholders through the Land for Wildlife program (see Chapter 3 and this chapter);

- the Nature Conservation Covenant Program (see Chapter 3);

- the Roadside Conservation Committee;

- Urban Nature, which supports established urban bushland conservation groups, as well as encouraging new urban bushland and wetland conservation initiatives, and helps private landowners to manage their urban bushland and wetland areas more effectively for nature conservation;

- State Salinity Strategy programs, including Crown reserves, natural diversity recovery catchments, oil mallee and wetland monitoring programs (see p. 187);

- New Native Vegetation Crops (Search Program);

- Conservation and Land Management Act Section 16 and 16A management agreements;

- the Gascoyne-Murchison Strategy, including the Ecosystem Management Understanding (EMU) program, which aims to introduce pastoralists to ecological management of landscape and habitats by learning to recognise landscape processes, condition and trends;

- liaison with and advice to Regional Natural Resource Management Groups;

- Regional Community Herbaria, in association with CALM's WA Herbarium;

- the Vegetation Health Service;

- the Marine Community Monitoring Program;

- the Monitoring River Health program;

- Indigenous Heritage Unit advisory, education and training services and cooperative management programs;

- cooperative programs with external organisations, including the World Wide Fund for Nature Australia, Greening Australia WA, CSIRO and Land and Water Australia; and

- CALM's volunteers program.

Recovery teams play a vital role in the conservation of threatened species and ecological communities. Most recovery teams include local people, often representing local community groups, who can work in partnership with government staff and scientists to ensure effective conservation work.

Consultation

Community support for biodiversity conservation programs is greatly enhanced when effective consultation occurs before decisions are taken. Already, CALM consults widely when

Above CALM technical officer Colin Ward works alongside a paying volunteer during a *LANDSCOPE* Expedition to Peron Peninsula in Shark Bay. Such partnerships with the community are vital to advance biodiversity conservation.

Photo – Keith Morris/CALM

developing recovery plans for threatened species and ecological communities. Consultation during development of park and reserve management plans is routine, as is a public review period for draft plans. Unfortunately, due to the size of the task compared with the staff and other resources available, it has not been possible to include a public review period for draft recovery plans and interim recovery plans. Hopefully, this will be possible in the future.

Education and information

Without education, people cannot fully appreciate why biodiversity conservation is so vital, nor can public support for biodiversity conservation, including the conservation of threatened species and ecological communities, fully develop. Without readily available, accurate and up-to-date information, people cannot understand what needs to be done. CALM produces a wide variety of educational information, from schools projects on *Western Shield* to *LANDSCOPE* magazine, the NatureBase internet site and books such as this one. Educational material about biodiversity and its conservation is not limited, and should not be limited, to government— many other institutions, groups and individuals can and do contribute. The Department's goal is to continue to prepare and disseminate high quality, accurate information about the State's biodiversity and its conservation, so everyone can understand what actions are required, and what is actually happening.

Resources

The natural world is so varied and fascinating that wildlife television documentaries can draw audiences of many millions of viewers. Nature-based tourism is booming and brings considerable economic benefits to Australia. Many visitors to our country and State say that they would not have come here if it had not been for Australia's wild places and unique plants and animals. Most Western Australians travel to see the State's natural wonders and wildlife. Without proper protection, conservation and management of the State's wild places—and our amazing plants and animals—these attractions will degrade. Without careful, scientific biogeographic surveys, taxonomic research, studies of conservation biology, careful management of conservation reserves and other lands and waters, and appropriate recovery actions for threatened species and ecological communities, we will lose significant components of our biodiversity. We humans will be the poorer if this occurs.

Western Australia encompasses a vast, biodiverse land area, and has rich and varied marine habitats. While we have a relatively low human population for such a large area, we are able to impact on the environment in ways undreamt of even a few decades ago. Fortunately, we are also a relatively educated and wealthy society, capable of solving current environmental problems and preventing new ones from emerging. To achieve better biodiversity conservation, we must ensure that a significant proportion of public, business and private funds are allocated to conservation, and we must take personal responsibility for conservation in our daily lives.

References

Abbott, I. and Burrows, N. (2003). *Fire in ecosystems of south-west Western Australia: impacts and management.* Backhuys, Leiden, Netherlands.

Burrows, N. (2003). Fire for life. *LANDSCOPE* 18(4), 21–26.

Sattler, P. and Creighton, C. (eds) (2002). *Australian terrestrial biodiversity assessment 2002.* National Land and Water Resources Audit, Canberra.

Wallace, K.J. (compiler) (2001). State salinity action plan 1996. Review of the Department of Conservation and Land Management's programs January 1997 to June 2000. Department of Conservation and Land Management, Kensington.
http://www.naturebase.net/projects/pdf_files/salinity_report_june2001.pdf

Index of animals and threatened ecological communities

Page numbers shown in bold are the main entry for threatened or extinct species or threatened ecological communities. Common names are used for animals where available. Species with more than one common name may be listed more than once. Some other native species and feral animals are included. Species with no page number in bold are not listed as threatened in Western Australia.

Opposite The Vulnerable threatened ecological community at Roebuck Bay.   Photo – Jan van de Kam

2003238-1104-3M